Analog Circuit Design using Current-Mode Techniques

This book deals with the design of CMOS compatible analog circuits using current mode techniques. The chapters are organized in order of growing circuit complexity. The area of analog signal processing is introduced to readers as an evergreen subject of academics and research interest. The contents cover various interfacing circuits, different types of amplifiers, single-time constant networks and higher order networks for system design applications.

Features:

- Presents the design of CMOS analog circuits using the current-mode building blocks in a comprehensive manner
- Covers several amplifiers, different types of current mode filters including electronically tuneable ones with ease of integration features
- Discusses in detail the waveform generation circuits and their applications in communication systems
- Presents advanced topics related to field programmable analog arrays
- Proposes new current-mode activation function circuit for neural networks

This book covers electronic tuning aspects of circuits with the help of solved examples and unsolved exercises. The contents include many non-linear applications using current-mode techniques. In form of signal generators, many oscillators for various communication and instrumentation systems are presented. Few current-mode configurable analog cells and their tuning aspects are covered. Some SPICE based results are given in support of presented circuits. Each chapter discusses the IC compatibility issue, which provides useful direction for carrying out laboratory exercises on the subject. The book is expected to serve as an ideal reference text for research, senior undergraduate and graduate students in the field of electrical, electronics, instrumentation and communications engineering.

Analog Circuit Design using Current-Mode Techniques

Sudhanshu Maheshwari

CRC Press
Taylor & Francis Group
Boca Raton London New York

CRC Press is an imprint of the
Taylor & Francis Group, an **informa** business

First edition published 2023
by CRC Press
6000 Broken Sound Parkway NW, Suite 300, Boca Raton, FL 33487-2742

and by CRC Press
4 Park Square, Milton Park, Abingdon, Oxon, OX14 4RN

CRC Press is an imprint of Taylor & Francis Group, LLC

Library of Congress Cataloging-in-Publication Data
Names: Maheshwari, Sudhanshu, author.
Title: Analog circuit design using current-mode techniques / Sudhanshu Maheshwari.
Description: First edition. | Boca Raton : CRC Press, 2023. | Includes bibliographical references.
Identifiers: LCCN 2022059772 (print) | LCCN 2022059773 (ebook) | ISBN 9781032393070 (hardback) | ISBN 9781032516134 (paperback) | ISBN 9781003403111 (ebook)
Subjects: LCSH: Analog integrated circuits.
Classification: LCC TK7874 .M224 2023 (print) | LCC TK7874 (ebook) | DDC 621.3815--dc23/eng/20230406
LC record available at https://lccn.loc.gov/2022059772
LC ebook record available at https://lccn.loc.gov/2022059773

ISBN: 978-1-032-39307-0 (hbk)
ISBN: 978-1-032-51613-4 (pbk)
ISBN: 978-1-003-40311-1 (ebk)

DOI: 10.1201/9781003403111

Typeset in Sabon
by SPi Technologies India Pvt Ltd (Straive)

Contents

Preface

Scientific and technological knowledge has been a systematic evolutionary creation of human race. This has been a slow, need-driven process which involves the inquisitiveness of human race, thanks to their own creation as the most privileged species on Earth by supreme God. Out of all existing branches of exploration, the field of Electronics and Communication Engineering has become a mainstream of knowledge exploration and a medium of providing luxuries to mankind. This is now a well-established, yet continuously evolving, area of Engineering and Technology. One of the important subjects of this field is related to the naturally occurring signals: their handling, processing and interpretation. The naturally occurring signals are continuously varying variety and are called analog signals, their handling, processing and interpretation is referred to as analog signal processing. The operation on analog signals with current as main parameter of interest falls under the current-mode signal processing. There has been a tremendous impact of current-mode analog circuit design, and its system-level applications over last few decades. The author was first introduced to this area in 1998 by his Master's project and dissertation supervisor, Professor Iqbal Ahmad Khan, in Aligarh Muslim University. The first course on current-mode circuits and applications was introduced in 2007 at the Master's level in the Department of Electronics Engineering of the University. The course has been running since then and being delivered by the author. However, there is a scarcity of dedicated text books and reference books on the subject. With the author's interest and experience in teaching courses on analog circuit design, it was realized that there is a strong need to educate the engineering students on design of analog circuits using current-mode approach. This book is motivated by the fact and attempts to target undergraduates, post-graduates and research professionals in Electronics and Communication Engineering. The contents are gradually presented for beginners in the subject at the undergraduate level, with growing level of subject knowledge in subsequent chapters apt for the post-graduate level, to an in-depth sight into the area in later chapters, especially for research scholars and industry professionals.

With the introduction to subject area, the chapters take the readers to a systematic journey of analog signal processing in order of growing complexity of circuits. Chapter 1 presents a general introduction, where various aspects of analog circuit design and current-mode basics are described. Chapter 2 is related to the introduction of various current-mode analog building blocks. The CMOS circuitry of various building blocks are discussed. Chapter 3 covers analog current-mode interfacing circuits like voltage to current and current to voltage converters. Voltage and current buffers are further described. The amplifiers of different types, including summing/differencing, and instrumentation using current conveyors are presented. Chapter 4 covers the single time constant circuits realized using current-mode approach. The circuits covered in the chapter are integrators, differentiators and first order filters. Chapter 5 is devoted to the study of second order and higher order filters. Higher order networks are presented, with growing complexity for system design applications. The coverage includes both active-RC and active-C current-mode filters. The electronic tuning aspects of filter circuits are explained with several examples. The non-linear applications using current-mode techniques are subsequently presented in Chapter 6. The circuits presented are comparators, rectifiers, detectors and modulators. The signal generation for various communication and instrumentation systems is introduced with a variety of circuits, exhibiting distinct features in Chapter 7. The advanced topic related to system design is presented as Chapter 8. It covers an important component of field programmable analog arrays. Thus, the example of configurable analog block is presented. The actual simulation results of a CAB are included. The tuning and integration aspects are further discussed. The challenges, future directions along with required research initiatives are explored in Chapter 9. A novel application of current-mode techniques in realization of an activation function for neural applications is proposed in the chapter.

Each chapter is supplemented with solved and unsolved practice exercises, especially for undergraduate and post-graduate readers. Illustrative simulation examples are given for several circuits in order to motivate the readers. Each chapter covers the IC compatibility issue of presented contents, which provides useful direction for carrying out laboratory exercises on the subject. It is expected that the book would invoke interest to a broad spectrum of readers.

Sudhanshu Maheshwari
Professor, Department of Electronics Engineering
Aligarh Muslim University, India

Acknowledgement

The duties assigned by God are best to be carried out without waiting for the fruits, and this book is one of my duties with some fruitive outcome. My thoughts dwell on my late father (shri S. P. Maheshwari), who always dreamt of his son authoring a good book. Unfortunately, I missed out sharing about this project to my beloved mother, who departed just a month before this project was undertaken, in first quarter of 2022. Nonetheless, their blessings continue to shower on me. The project took off after some flattering words of admiration by my two sons, who thought me the right person to write! My better half's silent moral support, as ever, is worthy of some appreciation. I sincerely thank CRC Press and my commissioning editor, Gauravjeet Singh, and editorial assistant, Isha Ahuja, for arranging timely reviews and helpful advice throughout this writing journey. I am thankful to Dr Deepak Kumar Agrawal for some crispy drawings, which can be found in most of the chapters of this book. The topic of this book is closely related to the courses I deliver at various levels of Electronics Engineering programs and to my research area. Kushaagra Maheshwari (B. Tech. Electronics and Communication) deserves a worthy mention for some critical comments on organizing the contents' files. I extend a warm thanks to all who have helped me throughout the 26 years of my knowledge exploration mission. The book assumes this decent look thanks to the efforts of the editorial staff at CRC Press.

Sudhanshu Maheshwari

Author

Sudhanshu Maheshwari obtained his Bachelor's, Master's and Doctoral degrees in Electronics Engineering from Z. H. College of Engineering and Technology, Aligarh Muslim University (AMU). He is working as full Professor in the Department of Electronics Engineering, AMU, Aligarh, India. He has been associated with teaching and research for more than 27 years in the area of Electronic Circuits and System Design, with specialization in current-mode analog signal processing. The prestigious World's Top 2% List of Scientists by Elsevier and Stanford has been continuously including his name in the subject area for last three years (2022, 2021 and 2020). He has published more than 150 research papers across the leading indexed international journals, guided many Doctoral and a large number of Master's candidates in the area of current-mode circuit design. He is on the reviewer panel of a large number of circuits and system journals.

Chapter 1

Subject introduction

This chapter introduces the topic of analog signal processing in the context of the current-mode approach and its motivation and the prevailing bottlenecks with the conventional methods. The traditional methods of designing analog circuits are briefly considered qualitatively, followed by the need to employ an alternative approach, namely the current-mode approach. Thus, the chapter begins with an introduction to signals and the processing of signals. The processing of signals covers a general discussion on the various electronic functions encountered in the subject area. The use of discrete and integrated components is covered briefly, followed by the techniques employed for the processing of signals. A discussion on passive, active-RC, active-C and other techniques are deliberated upon. The circuit parameters or specifications required in analog circuit design are also discussed. The problems of voltage-mode operation are highlighted, leading the discussion to the need for current-mode techniques. The chapter prepares the readers to easily grasp the forthcoming chapters.

1.1 SIGNALS

The nature of real-world phenomena can be understood in terms of signals. The signals as they occur in nature are mostly analog, a mode which is easy to interpret by living species (mankind). The continuity of any such phenomenon is important for its interpretation by mankind. There are numerous examples of such analog signals. It can be a daily and round the clock temperature variation of a place, the change in humidity levels, the light intensity in a room, voice, images, the traffic conditions on a highway, the intensity of naturally occurring seismic vibrations and so on. As an example of analog signals, the temperature variation in a North Indian city in the month of January is shown in Figure 1.1. With the growing intervention of technology into the lives of mankind, the monitoring, interpretation, processing and transmission of most of the naturally occurring analog signals have become a well-defined area of study. The analog nature of signals poses a challenge owing to the unpredictability in terms of their values and changes

DOI: 10.1201/9781003403111-1

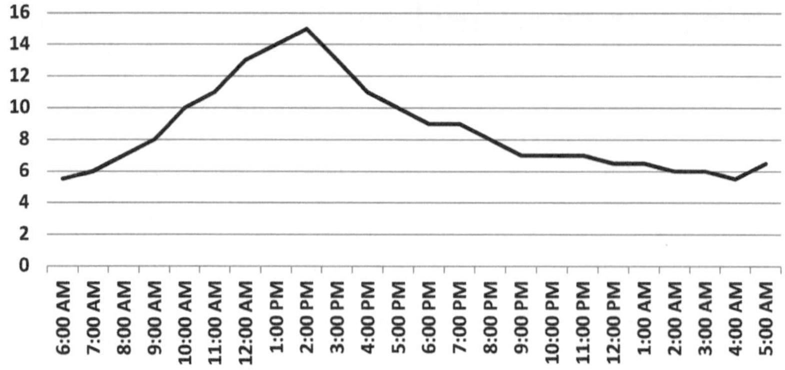

Figure 1.1 Typical analog signal example showing temperature (°C) variation with time for a North Indian city in January.

over time. Once an electrical/electronic means takes care of their conversion into the preferred mode, the next task necessitates further classification of these signals on the basis of their electrical identity. Thus, the signals can be of three main types: voltage, current and charge. Although purely for the sake of classification, these three types are still inter-related by electrical rules and theorems and so are inter-convertible as well. Out of the three mentioned types, the first one, namely the voltage signal, is the most readily understood and easy to interpret. This is because of the way electronics engineering has evolved over the decades, where the voltage signal has been the main parameter of concern, with the other two parameters (current and charge) being in secondary roles in most of the developments. Therefore, the voltage signals have been well understood and even monitored by available measuring instruments. Though the measurement of the other two is also well studied and available instruments and techniques cater to their measurement and monitoring as well, the perspective of the community studying the subject has always been inclined towards voltage signals only. However, there is scope for exploring the other types of signals. The signal to be of concern in the present study is the current signal. The talk of their interdependence and interconversion does exist. The pros and cons of the two types, namely, voltage and current, will also keep the debate on, as far as their comparisons are concerned. However, many positives would be highlighted and the benefits of the current signal would be explored once we progress into this study in subsequent sections and chapters. A representation of time-varying analog voltage and current signals is shown in Figures 1.2 and 1.3, respectively, for readers not familiar with the terminology.

Figure 1.2 shows a continuous time signal represented as a varying voltage with respect to time. Similarly, Figure 1.3 shows a continuous time signal represented as a varying current with respect to time.

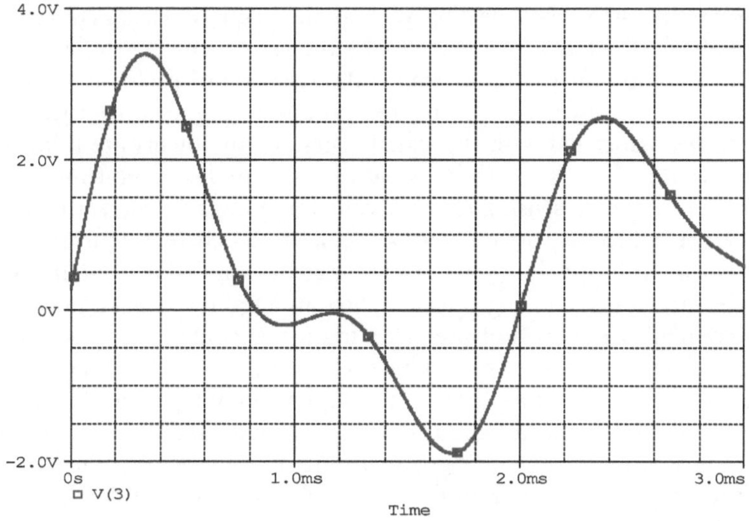

Figure 1.2 Typical time-varying voltage analog signal.

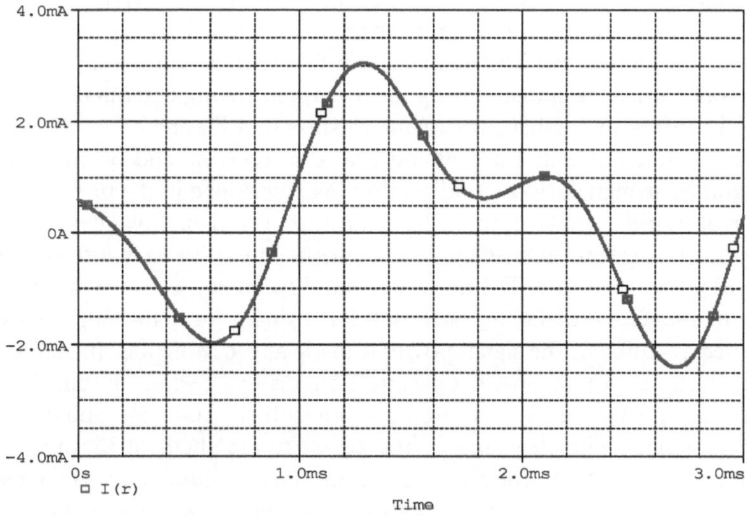

Figure 1.3 Typical time-varying current analog signal.

1.2 PROCESSING OF SIGNALS

The signals occurring in nature need to be converted to electrical/electronic equivalents, either as voltage or as current signals. The so obtained voltage or current signals require extensive processing before being actually used for either transmission or further interpretation. This is referred to as the pre-processing step when dealing with analog signals. The small or weak nature of such signals is one reason why the pre-processing step is needed. The other reason for pre-processing is the selection of preferred frequency content from the signal. It can further be due to the need to reject noisy fluctuations associated with the signal. These requirements are fulfilled by pre-processing elements, in the form of various simple components, devices or circuits. A simple sensing element called a voltage to current converter may be needed to convert the voltage signal to its current equivalent. The requirement to boost the signal level is met by employing a circuit which provides larger signal swings and is called an amplifier. The elimination of noisy content or selection of preferred frequency content is carried out by circuits which are referred to as filters. On a more complex side, the need to change the signal nature from analog to its digital equivalent necessitate analog to digital converter circuits. These elements and circuits employed for pre-processing of analog signals are often referred to as analog front end. The design of analog front end is a challenge for circuit designers. This is mainly due to the continuous time-varying nature of analog signals, their infinite number of possible values at different times and the requirements to be fulfilled for a given application. These are better covered under analog design trade-offs, which take into account numerous design parameters. The best possible optimum solution is then obtained. This aspect would be dealt with in detail time and again within this text. As one example of pre-processing, a 4 KHz sinusoidal signal corrupted by high frequency noise is cleaned before any further processing, as shown in Figure 1.4.

Another stage of signal processing is at the receiving end of information after going through various conditioning. As mentioned in the preceding section, real-world signals need to be transmitted and finally interpreted by mankind. The restoration of signals in a form suited for easy interpretation requires post-processing. The need for digital to analog converters arises for restoring the nature of signals back to a form suited for mankind. The use of amplifiers to enhance the signal power is made at the receiving end. Similarly, the signal undergoes a similar operation for driving certain output devices, which can provide easy interpretation of information contained in the received signals. The cleansing of the signal to rid them of the unwanted fluctuations, distortions and frequency components again demand the use of appropriate frequency selective and distortion suppressing circuits. Therefore, the processing of signals, in general, requires numerous basic electronic functions to be realized. The list of such functions is exhaustive. Some examples include amplifiers, rectifiers, filters, data converters, oscillators, multipliers, detectors, comparators and many other operations. These functions find applications in processing, transmission, reception and interpretation of biomedical signals, audio signals, video signals and RF signals, and form the basic building blocks of an electronic/instrumentation/communication system. For illustration purposes, Figure 1.5 shows a 101 MHz signal selection from a band of frequency, which is another example of signal processing.

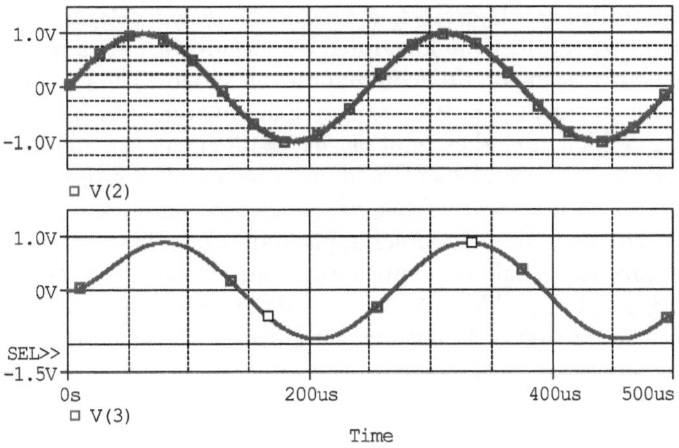

Figure 1.4 A 4 KHz signal corrupted by high frequency noise (above) after suppressing the noise.

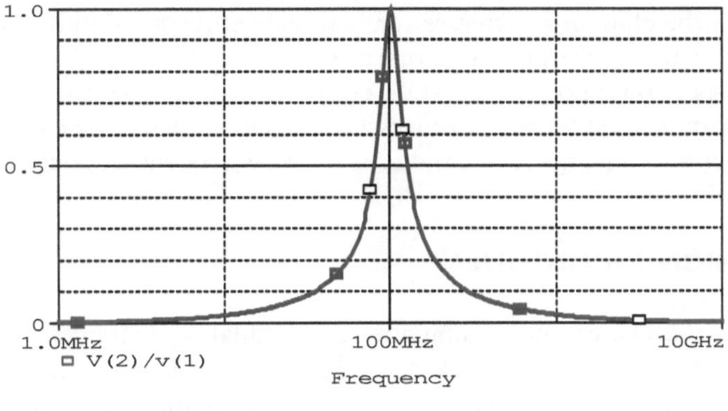

Figure 1.5 An illustration of signal processing showing a 101 MHz signal being selected with precise gain.

1.3 COMPONENTS

The electronic components, devices and integrated circuits available for performing various functions provide circuit designers with options to choose the optimum solution for a given problem. The components in the form of resistors, capacitors and inductors are the basic two terminal passive elements which find applications both in combinations and in conjunction with other devices and/or integrated circuits. Devices like diodes, with all their variants, bipolar junction transistors and metal oxide field effect transistors form the backbone of electronics engineering. These find applications in the design of most of the circuits for analog signal processing. The integrated circuits, like operational amplifiers, voltage regulators, timers, voltage-controlled oscillators, etc., are useful for system design applications. The components, devices and integrated circuits together are a part of a typical electronic/instrumentation/communication system. The R, L, C passive elements are an essential part in designing attenuators, phase lead-lag networks and tank circuits; they fulfil biasing requirements of transistor-based circuits, act as coupling elements and are also useful at radio frequencies. Diodes play a crucial role as switches, rectifiers, limiters, protective elements, detectors and in many other applications. These are also extensively used in displays, optical communication, voltage regulation, etc. The two transistor types, BJT and MOSFET, are used as amplifiers, switches and buffers, and in the design of multistage amplifiers and oscillators. Transistors also find applications in the design of integrated circuits. In a nutshell, the components, devices and integrated circuits are required for the processing of signals and hence in design of circuits and systems. From the integration viewpoint, most of the mentioned components and devices have been made a part of the integrated circuit (IC). Inductors with very low values are only feasible, otherwise they are the most bulky element to integrate. The integration of resistors and capacitors is also limited with respect to achievable values, as the chip area increases for high valued resistors and capacitors. The same is true for large transistors as well. For instance, a MOSFET with a large aspect-ratio poses integration challenges. The above-discussed components can be used for signal processing and circuit design by employing a number of techniques, which are discussed in subsequent sections.

1.4 DESIGN TECHNIQUES

Based on the choice and combinations of available components, devices and integrated circuits, there are several possible solutions to the problem of signal processing. The discrete and integrated solutions are two broadly classified methods. The discrete solution employs the components as available in discrete forms. These are mainly resistors, capacitors, diodes, transistors and integrated circuits. The undergraduate conduction of lab courses relies on using these components for solving simple problems in signal processing. For example, the combination of diodes, resistors and capacitors find applications in the processing of signals, for example, rectification, clipping, clamping. The collection of discrete transistors, resistors, capacitors is useful in designing circuits for amplification, signal generation, etc. The use of off-the-shelf integrated circuits, like operational amplifier, along with resistors, capacitors and diodes, serves the need for amplification, filtering, signal generation, multiplication, precision rectification, detection, etc. However, the integrated approach is mainly useful for imparting actual design and fabrication know-how to undergraduate students. This approach uses transistors (MOSFETs or BJTs) and other components and is realized using the same technology, with the aim to fabricate the IC.

The available techniques of analog signal processing are as follows: active-RC, active-R, active-C and active-only. The active-RC technique is based on the use of active element, R and C components. The active-R technique relies on the use of active element along with only resistors as the passive component. The active-C technique is based on the use of active element and capacitors as the only passive components. The active-only technique is different from the others in employing active elements only. Each of the technique is briefly discussed. The active-RC technique is mainly used to realize filters and oscillator circuits, where the pole-frequency, quality factor, bandwidth and oscillator frequency are some of the circuit parameters. The parameters depend on RC time constants and are normally varied by using variable resistors. The drawback of this technique is the fixed nature of time constants, which cannot be varied electronically. The use of variable resistor is not the preferred choice to vary a circuit parameter. A solution to this problem is found in active-C techniques. The resistors are replaced by resistor-equivalents, in the form of either inherent resistive elements within the circuit or active resistor simulators based on transistors. The advantage of this technique lies in resistor elimination, which reduces chip area and circuit noise, which would be contributed by resistors. Another advantage of the active-C technique is the possibility of electronically tuned circuits, where the equivalent/simulated active resistors can be tuned through an external voltage or current. The other technique, namely active-R, eliminates capacitors from the realizations. Instead, the internal poles of the used active element are exploited to realize time constants. The time constants depend on the external resistors and the internal pole-forming capacitors, rather than on physical capacitors. This technique has limited applications, as the internal poles restrict the available time constants, hence restricting on achievable circuit parameters. The active-only technique uses the active elements only. The resistive and capacitive effects to realize the desired time constants are obtained from circuit internal poles only. However, the technique has serious limitations with respect to its possible applications. This is due to its sole dependence on active elements' inherent poles. Most of these techniques employ operational amplifiers or operational transconductance amplifiers as the active element, besides external resistors and/or capacitors. Another technique which is especially used for sampled data systems is the switched-capacitor technique. It is mainly targeted for designing discrete-time analog circuits. The aim of such a design is the integration of a complete circuit, rather than discrete assembly. The components used are mainly CMOS operational amplifiers and switches, besides capacitors. All of these techniques are well understood for their operation with voltage as input and output signals. There is a lot of text available on the subject which cover these topics. However, the operability of circuits with current as input and output signals is not covered well in the available text. The motivation for such an operation is also missing, which is important for undergraduate/post graduate and research students. The limitations of existing techniques with above-mentioned components must also be emphasized before actual coverage of this topic. The following sections take the readers to explore all these aspects.

1.5 CIRCUIT PARAMETERS

The design of analog circuits relies on accurate realization of certain parameters, such as the resistor ratio, capacitor ratio, resistor-capacitor products, inductor-capacitor products, ratio of transconductances, etc. The ratio of resistors or transconductances is desired for amplifier design. The resistor-capacitor products (time constants) are used for designing integrators, differentiators, frequency selective networks, signal generators, etc. The inductor-capacitor product finds use in the design of tuned circuits and signal generation. The capacitor ratio mainly finds applications in the design of sampled data circuits, where the switched-capacitor technique is employed. Now let us come to the actual design parameters for a typical analog circuit used in analog signal processing. The design of analog circuits is tricky as well as intuitive because of the number of such parameters, which are too many even for a simple circuit. For example, an amplifier is characterized by voltage gain, input impedance, output impedance, dynamic range, bandwidth, slew-rate, power dissipation and noise. A filtering circuit is characterized by passband gain, stopband attenuation, pole-frequency, bandwidth and quality factor. Similarly, other signal processing functions can further be studied for their design parameters. These parameters are also referred to as specifications. It is difficult to fulfil all the design specifications. Therefore, certain tolerances are always mentioned for a given desired specification. The parameter without any specified tolerance becomes the primary one, which has to be met with good accuracy. Otherwise, a trade-off is the best choice when fulfilling the design specifications. As an example, an amplifier with constant gain-bandwidth product compromises one of the two parameters, namely gain or bandwidth, when fulfilling the other parameter. As another example of amplifier design, the slew-rate and power budget require a trade-off, both being a function of biasing current. The dynamic range and the supply voltage are also related parameters, the former being a function of the latter. Supply voltage scaling results in lowering the dynamic range of the circuit. The signal swings, which directly refer to dynamic range, are also limited by circuit noise figures. For example, an amplifier with larger noise figure would restrict the minimum signal that can be faithfully processed by the amplifier. In a nutshell, it is to be emphasized that there is trade-off in the design of analog signal processing circuits. The role of an analog designer is thus very crucial in the sense that, besides following the circuit laws, there is an element of intuition for the purpose. The role of intuition is in designing circuit topologies which are not sensitive to various undesired circuit effects. These undesired circuit effects arise from tolerances and parasitics. The circuit design parameters depend on the non-idealities of the components. The passive elements exhibit tolerances and parasitics, while active elements show parasitic and non-ideal effects. The parasitic effects degrade the frequency performance of circuits. A good design can ensure insensitivity to these undesired effects, besides fulfilling the circuit specifications. The optimum solution to the problem of analog signal processing is found in terms of components' requirement and a circuit topology that closely meets the desired specifications with tolerable accuracies. There are dedicated texts which deal with the design of analog circuits using CMOS technology [1, 2]. In order to provide a feasible and optimum solution to analog signal processing problems with improved performances, the use of current-mode techniques is to be explored in the subsequent section.

1.6 NEED FOR CURRENT MODE

The problems with the voltage-mode operation are to be identified before exploring the alternative design approach, namely, the current-mode approach. It is well known to readers of the subject that the measurement of voltages is quite easy. The easy availability of voltmeters, multimeters and oscilloscopes at our homes, offices and laboratories enables us to measure the ac supply voltage, electronics gadgets' voltage and experimental circuits' node voltages. Whereas, the measurement of current is not as easy and hence not common in day-to-day life. The easy measurement of voltage signals through oscilloscopes in the working environment (academics and research) keeps us biased in favour of this parameter, often ignoring the significance of current. Moreover, many of the early developed integrated circuits, like operational amplifiers, are based on voltage input, voltage output theory. The example of operational amplifier is important because it was one of the first few ICs that brought about a revolutionary change in analog signal processing. This biased opinion in favour of 'voltage' was proved wrong once the concept of current-mode became known. The gain-bandwidth product limitation of voltage input, voltage output operational amplifier puts a restriction on the useable bandwidth, hence limiting the high frequency performance of operational amplifier-based realizations. The early operational amplifiers with large power supply operation did not pose noise problems, which became serious with scaling of supply voltages for low voltage, portable systems. Basic operations like summing and differencing of signals with opamp require the use of resistive elements with appropriate matching. The parasitic capacitances associated with opamp circuits further restrict their high frequency applications. The limited slew rates of voltage operational amplifiers are not suited for analog signal processing, demanding high speed operation. All these limitations led to the motivation to find an alternative approach to analog signal processing. This was found in the current-mode approach. The use of 'current' as an active parameter instead of voltage became the design tool. The advantages to be explored were in terms of low supply voltage operation, improved dynamic range, higher bandwidths, greater slew rates, ease of performing simple operation like summing and differencing, reduction of noise, etc. Thus emerged the concept of current-mode circuits, which started finding applications in analog signal processing. Two types of current-mode definitions emerged: one that says that a circuit with current input and current output qualifies to be called a current-mode circuit, and the other that says that a circuit designed using a current-mode approach/active element qualifies to be called a current-mode circuit. An extension of the two says that a circuit designed using a current-mode element which accepts a current input signal and provides a current output signal also falls in the class of current-mode circuits. All the three mentioned definitions are acceptable versions of a genuine current-mode circuit. In order to introduce the concept to the readers who are at the half-way mark in their undergraduate Engineering programmes, the topic of current mirrors is not new. It becomes the starting point for further study into the current-mode circuits. The current mirror forms a basic building block of many integrated circuits. It is used for biasing of ICs and as active load for many amplifiers. The current mirror is one of the simplest possible circuits, which is studied as a current input and current output function. Although the input current is often shown to be derived from a voltage source, in this case, it is referred to as a current source circuit. Both the transistor types, namely MOSFET and BJT, are used for the design of current source and current mirrors. The desired characteristics of a current mirror are low input impedance, high output impedance and constant current gain, often set to unity, with design flexibility to adjust the current gain. The input and output impedance requirements as above are desirable for avoiding loading problem. Such a circuit with current input and current output would not load the signal source and the output can drive the load circuit without attenuation. The basics and intricacies of using the current-mode approach can be found in a text which appeared long back [3]. With this first-hand introduction to the topic, its need and benefits, the following chapter is exclusively devoted to an in-depth study of some basic current-mode building blocks.

1.7 SUMMARIZED CONCLUSION

This introductory chapter of the book lays the foundation for the subsequent chapters to follow. A general discussion of signals is provided, and the acceptance of current signals for analog signal processing is emphasized. The processing of signals and the components used for processing are introduced. The design techniques for analog signal processing are deliberated upon. The readers are introduced to the various circuit parameters encountered in analog circuit design. Finally, the need for current-mode techniques for analog circuit design is explained. It is realized that the use of current-mode techniques will benefit the design of future electronic and communication systems.

Chapter 2

Current-mode blocks

This chapter introduces the concept of current-mode analog building blocks. The introduction of current conveyor at the time when operational amplifier was getting popular is also discussed. A brief mention of the operational transconductance amplifier and its comparison with operational amplifier is made. The superiority of the current conveyor over the other two blocks is highlighted. The black box representation, port relationship, technology (CMOS or bipolar) used for implementation and description of current-mode building blocks are also presented. The presented building blocks' salient features are described and the different blocks are compared for their characteristics and performances. The building blocks to be presented are single-ended input and/or output, differential input/output current conveyors and electronically tuneable types. Therefore, the CCII, CCCII, EXCCII, EXCCCII and DVCC blocks are introduced to the first-time readers. The IC compatibility of such blocks is given, with specific realizations for various building blocks, which provides to the readers a starting platform to perform laboratory exercises. The chapter provides a first-hand experience to the readers at various levels to get acquainted with the current-mode building blocks.

2.1 CURRENT CONVEYOR

The time when operational amplifier was cementing its place as a standard analog chip for signal processing, the advent of a new active building block laid the foundation for a new era in analog signal processing. It was in the late sixties that Sedra and Smith introduced two building blocks and named these as first- and second-generation current conveyors. The generation naming was a convention followed by inventors, as the two different circuit topologies were designed in 1968 and 1970, respectively. These were referred to as CCI and CCII [4, 5], respectively. The general term 'current conveyor' emerged from the circuit functionality, which allowed an input current to be carried accurately at the output node, while also ensuring a voltage replication between the two input nodes. The definition of current

DOI: 10.1201/9781003403111-2

conveyor specifies that it is an accurately defined current follower, with the added advantage that voltage at one of the input terminals is also conveyed to the other input terminal. The CCI is not to be elaborated further, and the details to follow shall confine to the CCII. The reason for this biased favour for the latter is the general acceptability of this building block and its versatility for analog signal processing. The advent of CCII was important because of its superiority over the operational amplifier. The other active building block available at the time was the operational transconductance amplifier (OTA). The OTA is known to be a voltage input, current output building block. The operational amplifier's high input resistance, low output resistance, infinite open loop voltage gain and differential input features are known to the readers. The infinite input resistance, infinite output resistance, bias current controlled transconductance and differential input features of OTA are also well known to interested readers. The operational amplifier does not exhibit electronic control over its parameters, while the OTA exhibits electronic control over the transconductance parameter by the external bias current. A brief look at the symbols and basic equation of operational amplifier and OTA is in order. The symbols are shown in Figure 2.1.

The operational amplifier is defined by the following input–output relationship:

$$v_o = A_o v_{id} \tag{2.1}$$

The OTA is defined by the following relationship:

$$i_o = g_m v_{id} \tag{2.2}$$

In equation (2.2), the transconductance parameter is given as:

$$g_m = \frac{V_T}{2I_B} \tag{2.3}$$

Equation (2.3) relates the transconductance with bias current (I_B), and V_T is the thermal voltage, which is approximately 25 mV at room temperature.

Figure 2.1 (a) Opamp symbol; (b) OTA symbol.

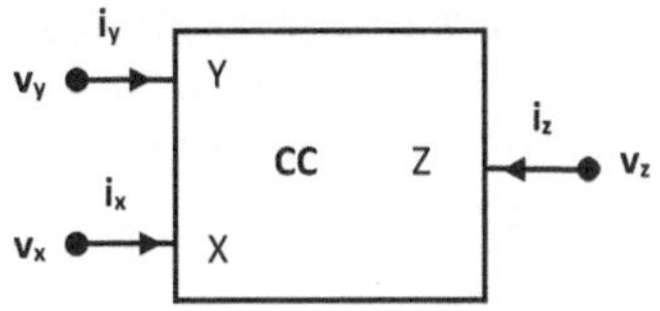

Figure 2.2 General symbol of a current conveyor (CC).

The electronic tuning property of OTA has no counterpart in the operational amplifier.

The symbol of a current conveyor is given in Figure 2.2.

The defining equation for the CC is given as:

$$i_y = gi_x, v_x = v_y, i_z = pi_x \quad (2.4)$$

The equation is valid for a CC of symbol as shown in Figure 2.2. The '*g*' value specifies the generation of CC, namely, first, second and third. The value of '*g*' for CCI, CCII and CCIII is 1, 0 and −1, respectively. Thus, for a CCI, the current at *Y terminal* follows the current at *X* terminal. For CCII, the current at *Y* terminal is zero. A CCIII has the current at *Y* terminal equal but opposite in direction to the *X*-terminal current. The node voltages are also shown at various terminals in Figure 2.2. The voltage following property from *Y* to *X* terminal is a common feature in all three generations. Similarly, the current conveying property from the *X* to the *Z* terminal is also common in all three types of CCs. The '*p*' in the current transfer expression of equation (2.4) signifies the direction of currents at *X* and *Z* terminals. The value of $p = 1$ for both *X* and *Z* terminal currents flowing either out of the black box or inwards into the black box. The value of $p = -1$, if one of the two currents flows inwards, while the other flows outwards, and vice versa. The value of '*p*' further classifies the CC into two types, in each of the three generations. These are referred to as CC+ and CC−, respectively. The CC types to be further discussed in detail in the next sections are CCII+ and CCII−.

2.2 SECOND-GENERATION CURRENT CONVEYOR

The CCII introduced in the previous section is the most versatile of the three generations. This is due to the advantage of a high impedance input terminal (*Y*) in CCII. The current at the *Y* terminal is zero, which makes the CCII *Y*-input ideal for voltage signals. The high input resistance at this node makes CCII different from the other two generations (CCI and CCIII). This property is similar to the input(s) of an operational amplifier, as well as those of an operational transconductance amplifier. All the three active building blocks thus have a common feature of a high input impedance voltage input terminal. The CCII (like the other two generations) has a current output at high impedance. This property is common to the operational transconductance amplifier. However, there is no counterpart in the operational amplifier. The current following property of CCII (from *X to Z*) has no analogy in either the operational amplifier or the operational transconductance amplifier. Similarly, the voltage following property at two input terminals (*Y* to *X)* has no analogy in either of the two available active building blocks. But this property is conditional in the operational amplifier, where a virtual short concept under negative feedback may be seen analogous to CCII voltage buffering property.

The second-generation current conveyor's ideal describing equation is reproduced below for further explanation.

$$i_y = 0, v_x = v_y, i_z = pi_x \tag{2.5}$$

Equation (2.5) can be further defined for a CCII+ and a CCII–, respectively, as expressed below.

$$i_y = 0, v_x = v_y, i_z = i_x \tag{2.6}$$

$$i_y = 0, v_x = v_y, i_z = -i_x \tag{2.7}$$

The ideal equivalent circuit of a CCII can be realized using a voltage controlled voltage source and two current controlled current sources, as is clear from equations (2.6 and 2.7). However, a more realistic model of a CCII is desired from the purpose of fully understanding the building block. Similar to an operational amplifier or an OTA, the CCII can be modelled using controlled sources and passive elements. The simplified model of a current conveyor is given in Figure 2.3.

The simplified model of the CCII as shown in Figure 2.3 includes an intrinsic resistance at the *X* terminal, which, as the name suggests, is inherent and hence not shown. It may be modelled as R_x. The two impedances at *Y* and *Z* terminal are the parasitic impedances, which are modelled as Z_Y and Z_Z, respectively. Each of the two includes a resistive and a capacitive component in parallel. The two can be expressed as below.

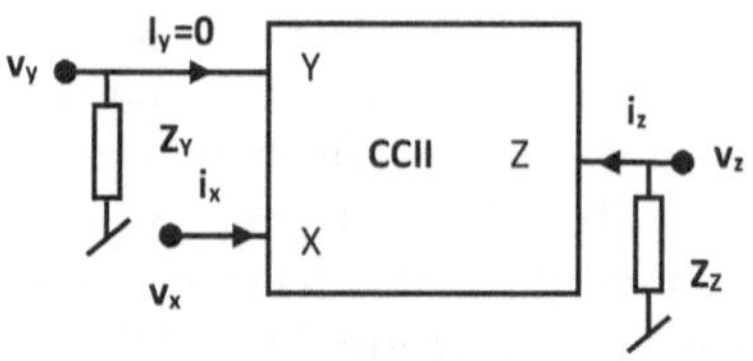

Figure 2.3 Simplified equivalent model of a current conveyor (CC).

$$Z_Y = R_Y // \left(\frac{1}{sC_Y}\right); Z_Z = R_Z // \left(\frac{1}{sC_Z}\right) \tag{2.8}$$

Alternatively, equation (2.8) can be expressed as follows:

$$Z_Y = \frac{R_Y}{sR_YC_Y + 1}; Z_Z = \frac{R_Z}{sR_ZC_Z + 1} \tag{2.9}$$

Equation (2.9) expresses the *Y* and *Z* nodes' impedance functions. The equivalent model of CCII can be interpreted in terms of the real behaviour of the active building block. The real CCII behaviour depends on the frequency of operation. The voltage following and current conveying properties are frequency dependent. The parasitic capacitance restricts the building block performance, which is decided by the actual parasitic values. The resistances at the *Y* and *Z* terminals, modelled as R_Y and R_Z, are ideally infinite. The small R_x is the input resistance at the *X* terminal, which is ideally zero. The precise values of these parasitic elements depend on technology in which CCII is implemented. The capacitances C_Y and C_Z refer to the *Y* and *Z* nodes, respectively. The non-ideal CCII defining equation is expressed as.

$$i_y = 0, v_x = \beta v_y, i_z = \pm \alpha i_x \tag{2.10}$$

Equation (2.10) suggests the voltage following action, with transfer gain as β, and current conveying action, with the transfer gain as α. The ± sign stands for a CCII±. The voltage and current transfer gains are shown to be real quantities. The frequency dependence of these gains can be better appreciated using a single-pole model. Therefore, the two transfer gain factors are expressed as below.

$$\alpha(s) = \frac{\alpha_o}{1 + j\omega/\omega_\alpha}; \beta(s) = \frac{\beta_o}{1 + j\omega/\omega_\beta} \tag{2.11}$$

In equation (2.11), the dc gains of the two transfer gain factors are α_o and β_o, respectively. It may be noted that CCII is designed for unity voltage and current transfer gains. The –3 dB frequencies of the two factors are ω_α and ω_β, respectively. These angular frequencies (radians/second) suggest the drop of dc gain by 3 dBs, at their values, in each case. The angular frequencies (ω_α and ω_β) are the actual pole frequencies for the voltage and current transfer gains. The voltage transfer gain pole frequency (f_β) and current transfer gain pole frequency (f_α) can be expressed as:

$$f_\beta = \frac{\omega_\beta}{2\pi}; f_\alpha = \frac{\omega_\alpha}{2\pi} \tag{2.12}$$

The unit for the pole frequencies in equation (2.12) is Hertz.

Practice Exercise 2.1: *What is the lower -3 dB frequency for the voltage and current transfer gains of a CCII if the angular frequencies are respectively 10E+7 rad./sec. and 12E+7 rad./sec.? (Hint: Use equation 2.11)*

Practice Exercise 2.2: *If the dc gains for voltage and current transfers of a CCII are unity, find the frequency at which the two gains drop to 0.8. (Hint: Use equation 2.11)*

Practice Exercise 2.3: *If a CCII has* $R_Y = 10\ M\Omega$, $R_Z = 5\ M\Omega$, $C_Y = 10\ pF$ *and* $C_Z = 5\ pF$, *find the parasitic impedance functions. (Hint: Use equations 2.8 and 2.9)*

After the block-level description of CCII, one possible CMOS circuitry for its implementation is shown in Figure 2.4 [6].

The CCII circuit in Figure 2.4 can be described by considering its constituent sub-blocks. The sub-blocks forming the complete circuit are (i) trans-linear loop, (ii) biasing network and (iii) current following sub-block. The trans-linear loop comprises four transistors (M1–M4) forming a closed loop, which is essentially a current-mode circuit. Transistors (M5–M7 and M10–M11) constitute the biasing network, along with the external current source I_o. The transistors M8–M9 and M12–M13 form the current following sub-block. The voltage transfer gain form the *Y* to the *X* terminal is

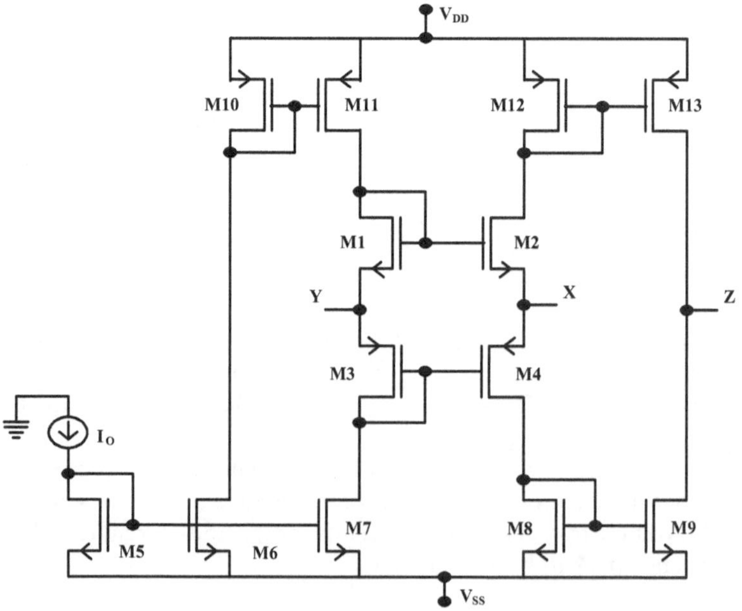

Figure 2.4 CMOS circuitry of CCII [6].

unity. The inherent X-terminal resistance for the trans-linear loop is given as:

$$R_x = \frac{1}{\sqrt{8\mu C_{ox}\left(\frac{W}{L}\right)I_o}} \tag{2.13}$$

Equation (2.13) assumes the matching of the transistors forming the loop (M1–M4) as:

$$\mu_n C_{ox}\frac{W}{L}_n = \mu_p C_{ox}\frac{W}{L}_p = \mu C_{ox}\frac{W}{L} \tag{2.14}$$

Equation (2.14) requires appropriate scaling of PMOS and NMOS transistors' as per the mobility factors of the two transistor types. The biasing network provides equal bias (I_o) to the trans-linear loop transistors M1 and M4, respectively, through the current mirror, namely M10–M11 and M5–M7 (current mirror along with steering transistor). The X-terminal current is conveyed to the Z terminal through mirrors M8–M9 and M12–M13.

2.3 EXTRA-X CURRENT CONVEYOR

The next active building block to be introduced is an extra X second-generation current conveyor, which is abbreviated as EXCCII. The EXCCII can be considered as a combination of two CCIIs. It incorporates an additional low input impedance X terminal, hence a high output impedance current terminal. The defining equations for an EXCCII are given as:

$$i_y = 0, v_{x1} = v_y, v_{x2} = v_y; i_{z1} = i_{x1}; i_{z2} = i_{x2} \tag{2.15}$$

The symbol of an EXCCII is shown in Figure 2.5.

If equation (2.15) is compared with the CCII defining equation (2.6), it is clear that an EXCCII is equivalent to two CCIIs of positive polarity, with a

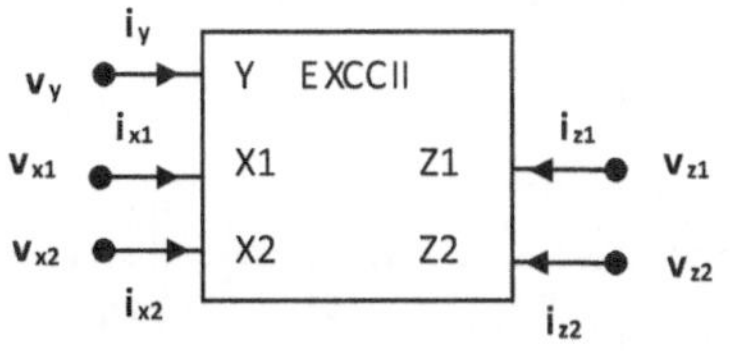

Figure 2.5 Symbol of an EXCCII.

common Y input terminal. The EXCCII active building block is thus more useful by incorporating two voltage followers at the input terminals. Rest of the theory related to the equivalent model, parasitic model and non-ideal behaviour for EXCCII is easy to follow from the earlier description of CCII. The only change is in terms of the two X stages and their own output Z stages. It is clear that the EXCCII is more versatile than a CCII, as it combines the features of two CCIIs.

Practice Exercise 2.4: *Obtain the equivalent model of an EXCCII using the model of a CCII (Figure 2.3).*
Practice Exercise 2.5: *Write the parasitic impedance expressions for an EXCCII using the CCII expressions of equations (2.8 and 2.9).*
Practice Exercise 2.6: *Write the non-ideal defining equations for an EXCCII. (Hint: Use equations 2.10 and 2.11).*

The CMOS circuit of an EXCCII is given in Figure 2.6 [7]. It is an extension of the CCII of Figure 2.4. An additional X stage (as X_2), and Z_2 is implemented in it.

The EXCCII comprises two input voltage followers and two current followers. For the beginners, it is interesting to develop a macromodel of an EXCCII, which can be easily simulated using freely available simulation

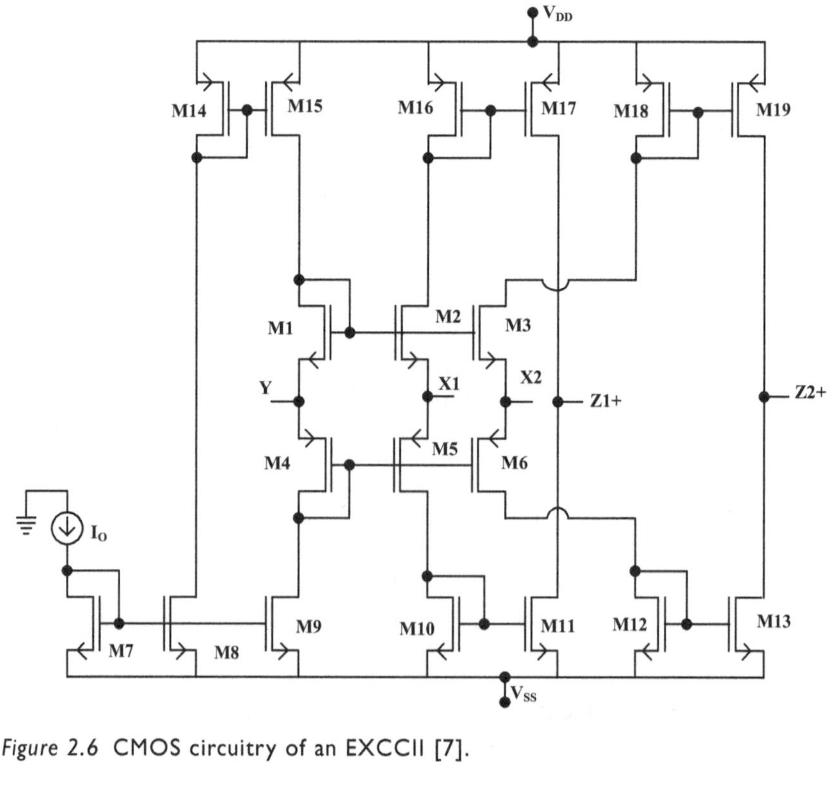

Figure 2.6 CMOS circuitry of an EXCCII [7].

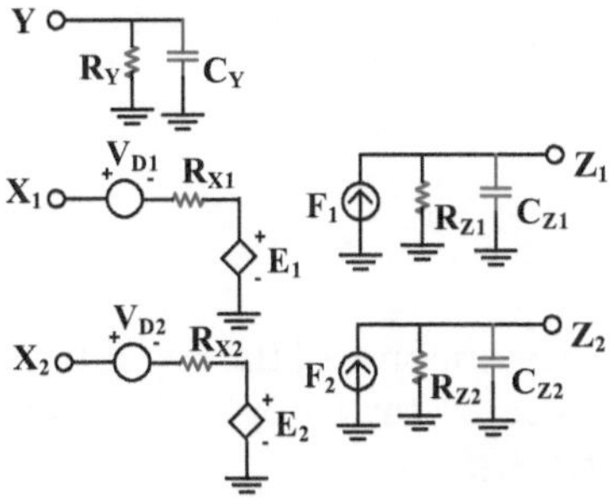

Figure 2.7 SPICE micromodel of an EXCCII.

tools, for example SPICE. This would provide an insight into the working of the building blocks, without going into transistor-specific details and would serve as a starting point to explore the marvels of this building block. The readers who are new to SPICE are required to refer to relevant texts and sources before attempting this exercise. The macro model for an EXCCII is shown in Figure 2.7.

PSPICE code:

```
**EXCCII macromodel for Electronics Engineering graduates
**Y = 1, X1 = 2, X2 = 3, Z1 = 4, Z2 = 5
**Y terminal impedance
RY 1 0 10MEG
CY 1 0 1.0p
**Voltage controlled voltage sources
EX1 6 0 1 0 1
EX2 7 0 1 0 1
**The X2 terminal resistance
RX1 2 8 100
**The Rx2 terminal resistance
RX2 3 9 100
**Dummy voltage sources
VD1 6 8 0
VD2 7 9 0
**Current controlled current sources
F1 0 4 VD1 1
F2 0 5 VD2 1
**Z terminals impedances
RZ1 4 0 1MEG
CZ1 4 0 1p
RZ2 5 0 1MEG
CZ2 5 0 1p
**Test input
VIN 1 0 AC 1 SIN(0 1 1MEG)
**Frequency sweep
.AC DEC 100 100 100MEG
**Transient analysis
.TRAN 0.1N 5U
```

```
CL1 2 0 1p
CL2 3 0 1p
RL1 4 0 1K
RL2 5 0 1K
.PROBE
.END
```

The results for frequency response of the EXCCII are given in Figures 2.8. The X terminals are loaded by capacitors, whereas the Y and Z terminals have been loaded by resistors.

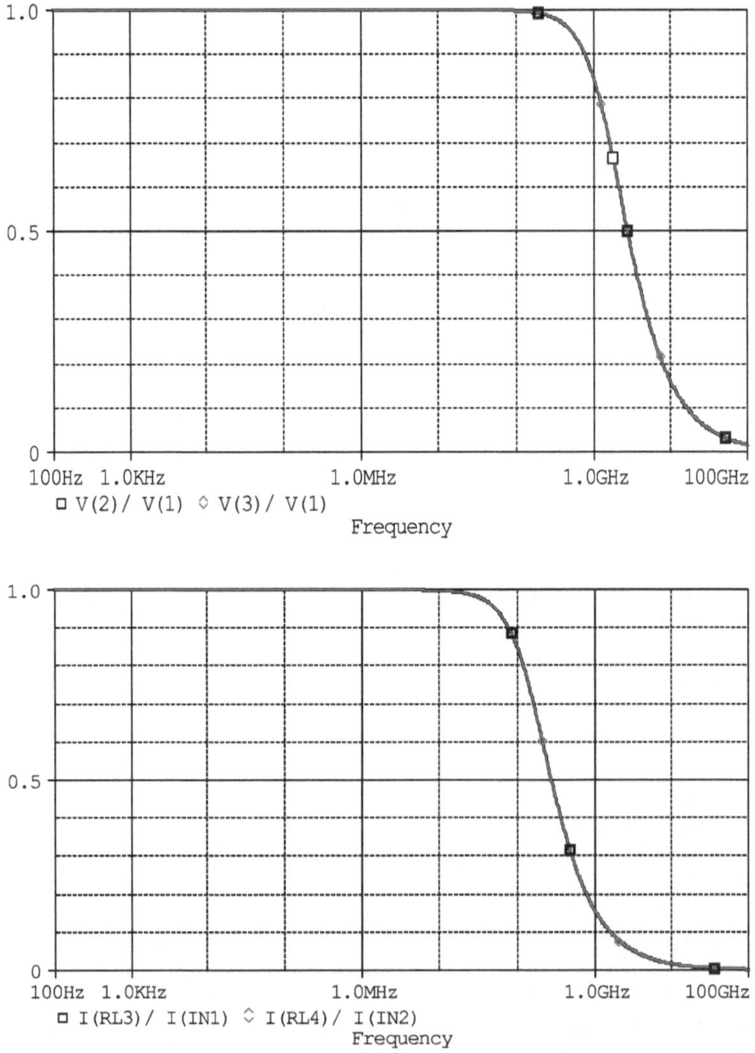

Figure 2.8 Voltage transfer gain plot and current transfer gain plots of EXCCII.

It is worth to note that like a CCII, an EXCCII can also be designed for negative current transfer gain from X to Z terminals. Moreover, an EXCCII can be designed for both positive and negative current transfer gains.

2.4 DIFFERENTIAL VOLTAGE CURRENT CONVEYOR

Both the CCII and the EXCCII introduced so far cannot process differential voltage signals. The need of differential signal processing arises from the noise and interference elimination point of view. It is a well-known fact that differential signal processing removes the common mode signals, hence improving the circuit performance. The differential amplifiers form the basic building blocks for many commercially available integrated circuits. The differential topologies also offer ease in integration of circuits for matching and symmetry reasons. The current-mode active building block which offers differential voltage signal processing is referred to as the differential voltage current conveyor, and is abbreviated as DVCC. The symbol of DVCC is shown in Figure 2.9. The defining equation for a DVCC is given as equation (2.16).

$$i_{y1} = i_{y2} = 0, v_x = v_{y1} - v_{y2}, i_z = i_x \tag{2.16}$$

The DVCC differs from a CCII by employing a differential voltage input stage. This input voltage is conveyed to the other input (X) at low impedance. The low impedance (X) input is ideal for current signal injection, and this current is conveyed to the high impedance output Z terminal. Like a CCII, the DVCC can be designed for negative current transfer gain from the X to the Z terminal. Moreover, DVCCs with both positive and negative transfer gains can also be designed. The CMOS circuit for implementing a DVCC is shown in Figure 2.10 [8].

It is interesting to note that a DVCC with grounded Y_2 terminal can operate as a CCII. Similarly, a DVCC with grounded Y_1 terminal yields $v_x = -v_{y2}$. Thus, an inverting voltage buffer operation is performed at the input terminals. This property is used to realize another type of CC known as inverting second-generation current conveyor (ICCII). An inverting CCII differs from a CCII in voltage transfer gain property, which is characterized by an inverting polarity, as compared to a CCII, which has a positive unity voltage transfer gain.

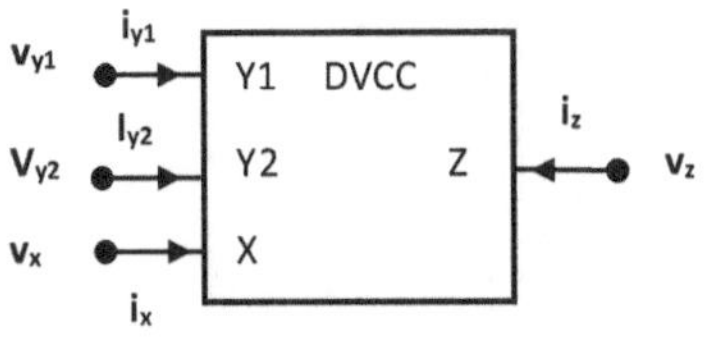

Figure 2.9 Symbol of DVCC.

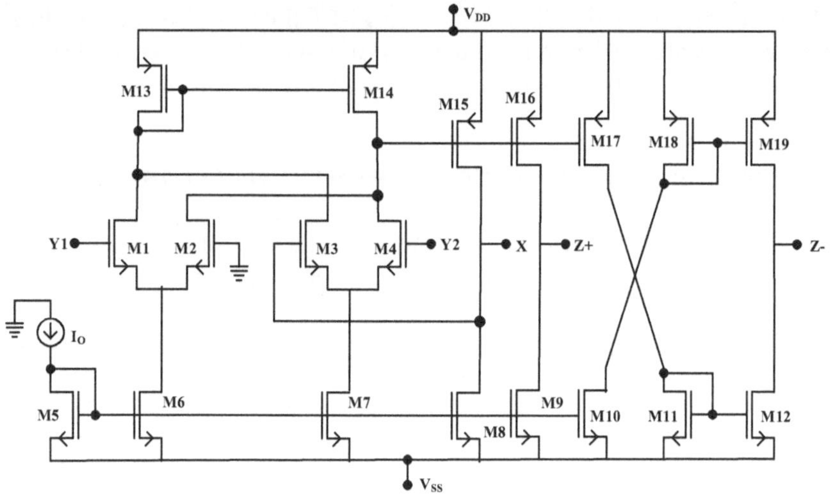

Figure 2.10 CMOS circuit for DVCC with Z+ and Z− stages [8].

2.5 CURRENT CONTROLLED CURRENT CONVEYORS

The CC types presented in the preceding sections fall in the category of non-tuneable active building blocks. There is no provision to vary a characteristic parameter electronically in all of these building blocks. This puts a limitation on their applications, where electronic tuning is important. It is worth noting that tuning is a mandatory feature for most of the analog signal processing circuits. The current controlled current conveyor (CCCII) is one such building block which allows electronic tuning of an intrinsic parameter [9]. The symbol of a CCCII is shown in Figure 2.11. It is characterized by the following relationship with reference to Figure 2.11.

$$i_y = 0, v_x = v_y + i_x R_x, i_z = i_x \tag{2.17}$$

Equation (2.17) suggests that CCCII is like a CCII, with the difference that the *X*-terminal intrinsic resistance is finite and electronically tuned by the external bias current. This property of a CCCII is like an OTA, where the

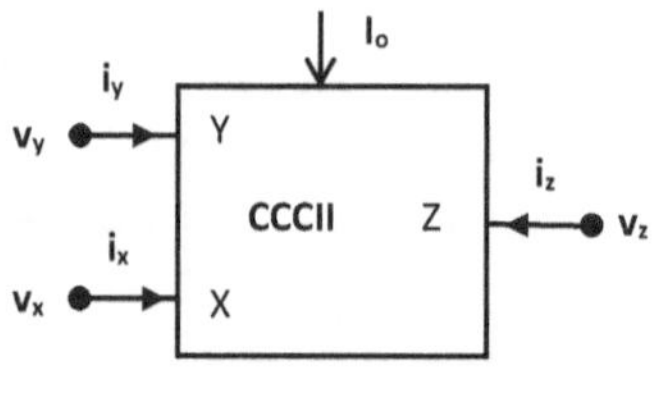

Figure 2.11 Symbol of a CCCII.

circuit transconductance parameter is electronically tuned by the external bias current. It is interesting to note that a CCCII can be treated as a non-ideal CCII with finite R_x which can be varied through an external bias current control. Earlier implementations of CCCII were made available in bipolar technology and found diverse usage for analog signal processing [9]. The CMOS circuit for realizing this block is same as shown in Figure 2.4. The value of R_x is repeated below for further explanation.

$$R_x = \frac{1}{\sqrt{8\mu C_{ox}\left(\frac{W}{L}\right)I_o}} \tag{2.18}$$

The CMOS circuit of Figure 2.4 is designed as a CCCII for variable I_o. The design for CCII is based on minimizing the R_x, whereas the design for a CCCII requires a variable R_x. Therefore, I_o is varied so as to obtain different values of R_x. This property of CCCII differentiates it from a CCII, where I_o is fixed at a large value to keep intrinsic resistance small. The electronic tuning enables CCCII-based analog signal processing circuits to eliminate resistive elements supposed to be present at the X terminal of CCCII. This allows for electronically tuneable realizations, which would be discussed in subsequent chapters. The CCCII are of different types based on the current transfer gains. For positive transfer gain, it is referred to as CCCII+. For negative transfer gain, it is referred to as a CCCII−. The circuit can also be designed for both positive and negative gain stages, where it is called a CCCII±. In fact, multioutput current conveyors only demand additional stages designed using current mirrors. Like a CCCII, an EXCCII can also be used as an EXCCCII (current controlled extra-X current conveyor), using the variable bias current (I_o), for obtaining different values of R_{x1} and R_{x2}, as per the relationship of equation (2.18), where both intrinsic resistance values are matched. The symbol of an EXCCCII is shown in Figure 2.12.

The defining equation for an EXCCCII is given in equations (2.19 and 2.20).

$$i_y = 0, v_{x1} = v_y + i_{x1}R_{x1}, v_{x2} = v_y + i_{x2}R_{x2}; i_{z1} = i_{x1}; i_{z2} = i_{x2} \tag{2.19}$$

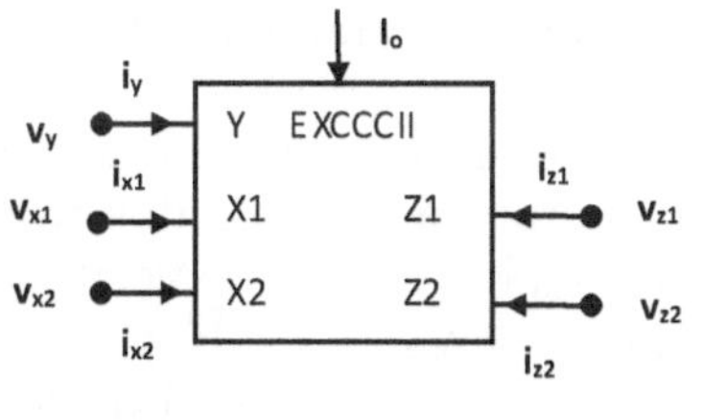

Figure 2.12 Symbol of an EXCCCII.

$$R_{x1} = R_{x2} = \frac{1}{\sqrt{8\mu C_{ox}\left(\frac{W}{L}\right)I_o}} \tag{2.20}$$

Practice Exercise 2.7: *What is the difference between a CCII and a CCCII?*
Practice Exercise 2.8: *Show the equivalent circuit for a CCCII.*
Practice Exercise 2.9: *What is the difference between a CCCII and an EXCCCII?*

2.6 OTHER CURRENT-MODE BUILDING BLOCKS

Besides the above-discussed current-mode active building blocks, there are several other available ones, which also find applications in analog signal processing. These are mainly variants of current conveyors only. The chief amongst these are current differencing buffered amplifier (CDBA), current differencing transconductance amplifier (CDTA), fully differential current conveyor (FDCCII) and dual X *c*urrent conveyor (DXCCII). The CDBA is a current differencing input block with a current output, with the additional feature of a buffered voltage output. The CDTA block has the current differencing input property, but the output stage incorporates a transconductance amplifier. The FDCCII has an interesting summing/differencing feature of input voltages, which are buffered, while current following stages are similar to a CCII. The DXCCII is the combination of a CCII and an ICCII (inverting current conveyor), with the advantage of voltage buffering action with both positive and negative transfer ratio. The tuneable ones include current controlled current differencing buffered amplifier (CCCDBA), current controlled current conveyor transconductance amplifier (CCCTA), etc. These are not to be further elaborated for now.

2.7 IC COMPATIBILITY OF BLOCKS

The current-mode active building blocks discussed in preceding sections are being seen as an alternative to conventional blocks available in the market. The conventional blocks available are mainly the operational amplifier and the operational transconductance amplifier. Both these blocks are commercially available as ICs in the market. The current-mode blocks are mainly targeted for complete integration of the circuit build around those blocks. Any analog signal processing application would be best suited for integration, to be part of a typical information/embedded system. However, in order to impart the know-how of these blocks at the undergraduate or postgraduate level, it is important to show their worth by using off-the-shelf chips. The testing of circuits using such active building blocks at the laboratory can be pursued as a part of practical courses. Therefore, the realization of such blocks is shown using commercially available chips. One such IC available is AD844, by Analog Devices. This is basically a current feedback operational amplifier. The CCII, EXCCII, DVCC, ICCII, DXCCII, etc., can be realized using this chip. The symbolic representation of a CFOA (AD-844 is one such IC), as shown in Figure 2.13.

The defining equation of a CFOA is given below.

$$i_y = 0, v_x = v_y, i_z = i_x; v_w = v_z \tag{2.21}$$

The CCII+ can be realized using a single CFOA (hence one AD844), with additional advantage of a buffered output in the form of *W*. This is evident from equation (2.21). Thus, a CCII-based circuit can be practically designed using a single AD844 IC. The CCII-type current conveyor requires two ICs; its realization is shown in Figure 2.14.

Practice Exercise 2.10: *Prove that the circuit in Figure 2.14 realizes a CCII with an additional buffered output.*

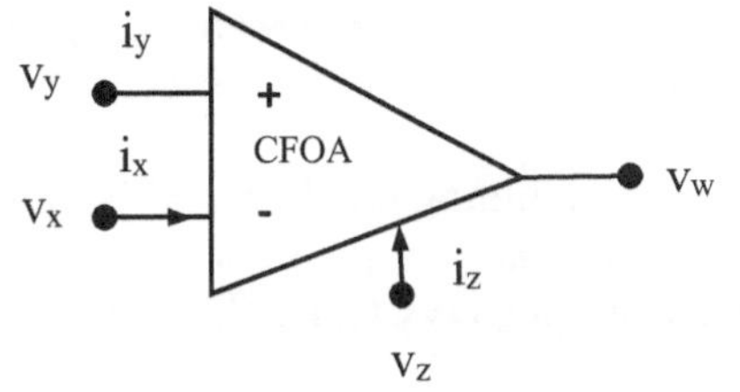

Figure 2.13 Symbol of a CFOA.

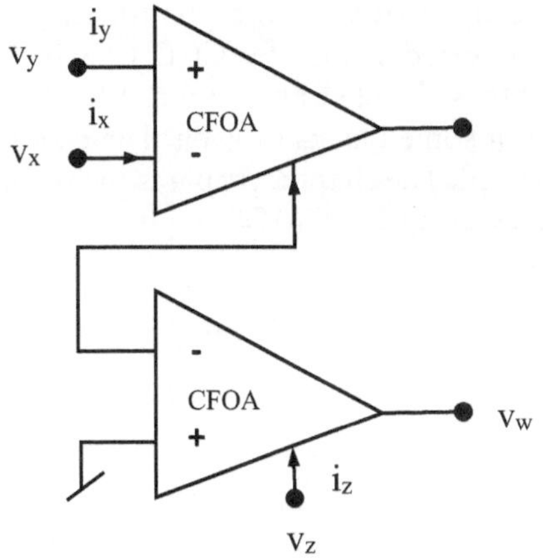

Figure 2.14 CCII realization using two ICs.

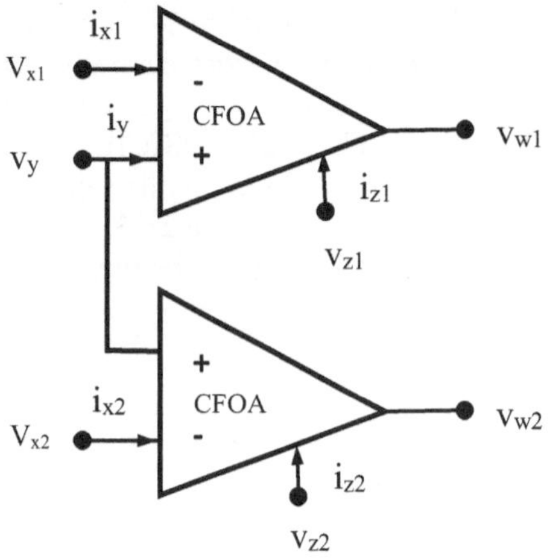

Figure 2.15 EXCCII realization using two ICs.

The EXCCII [10] can be realized using two AD844 ICs and its circuit is shown in Figure 2.15.

Practice Exercise 2.11: *Prove that the circuit in Figure 2.15 realizes the EXCCII with an additional buffered outputs.*

Rest of the current-mode building blocks are also compatible for AD844-based realizations. However, many of the remaining blocks require resistors in addition to ICs. For example, DVCC, ICCII, DXCCII are such blocks.

2.8 CONCLUDING REMARKS

This chapter introduced actual CMOS current-mode building blocks. The main building blocks covered include the CCII, CCCII, EXCCII and DVCC. The CMOS design of these blocks is presented. The SPICE netlist and characteristic verification of some blocks is given. The IC compatibility of various blocks is also covered. The chapter prepares the readers for undertaking actual design problems using the CMOS implementation of current-mode active building blocks.

Chapter 3

Analog interface and amplifier circuits

This chapter begins with some analog interface applications of the current-mode building blocks. The circuits presented are voltage and current buffers, voltage to current and current to voltage converters. Next types of circuits covered are amplifiers of single ended and differential nature. Single ended amplifiers and instrumentation amplifiers are introduced. The simple circuits covered form the core of the larger analog sub-systems to be realized in forthcoming chapters. The building blocks used for the purpose include CCIIs, CCCIIs and DVCCs. A critical discussion on input and output impedance requirements is provided for each of the presented circuits.

3.1 VOLTAGE BUFFER

The first circuit for interface applications using a CCII is a voltage buffer. It is required for impedance matching, while cascading sub-blocks for system design. It is also required for sensing the signal without loading the source. A voltage buffer is essential in driving the load circuit without signal attenuation. The current-mode approach to analog signal processing offers a simple voltage buffer in second-generation current conveyor, with accurate unity gain and ideal input and output impedances. The CCII-based voltage buffer (follower) is designed by using the voltage following action at the two input terminals of CCII. The simple connection to realize a CCII-based voltage buffer is shown in Figure 3.1.

Figure 3.1 realizes a voltage follower circuit with the following characteristics.

$$v_{out} = v_{in}; R_{in} = R_y; R_{out} = R_x \tag{3.1}$$

For a CCII, R_y and R_{out} are infinite and zero, respectively. These are the input and output resistances, respectively. Therefore, equation (3.1) can be written as below.

$$v_{out} = v_{in}; R_{in} = \infty; R_{out} = 0 \tag{3.2}$$

DOI: 10.1201/9781003403111-3

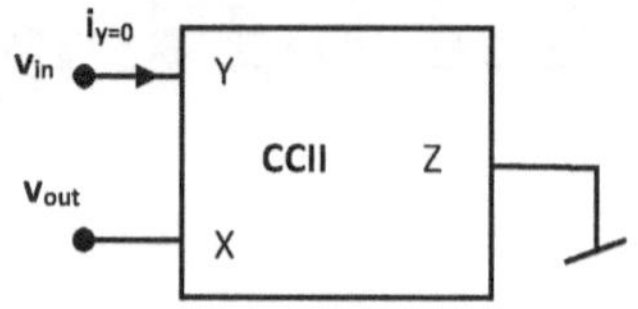

Figure 3.1 Simple voltage buffer using a CCII.

Equation (3.2) suggests that the impedance looking into the *Y* terminal is infinite and the output impedance at *X* terminal is zero. Thus, an ideal voltage follower can be realized using a CCII.

Practice Exercise 3.1: *(solved). A CCII with* $R_x = 100\ \Omega$ *is used as a voltage follower to drive resistive load* (R_L)*. What is the maximum value of load, if the output is desired to be at least 90% of the input signal?*

Solution: The output signal undergoes some attenuation due to R_x *between input and output. Once the load* R_L *is connected, the following current equation becomes valid.*

$$(v_{\text{in}} - v_{\text{out}})/R_x = v_{\text{out}}/R_L$$

$$v_{\text{out}} = [R_L/(R_L + R_x)]\, v_{\text{in}}$$

In order for the output to be at least 90% of the input, the following expression holds true.

$$R_L/(R_L + 100) = 0.9$$

The value of R_L *from the above is obtained as 900* Ω.

Practice Exercise 3.2: *What can be the maximum value of output resistance of CCII-based voltage buffer if it is to drive a load of 10 k*Ω*? The output is desired not to attenuate more than 5%.*

3.2 CURRENT BUFFER

The current buffer (follower) connection using a CCII is shown in Figure 3.2. The circuit characteristics are as follows.

$$i_{\text{out}} = i_{\text{in}}; R_{\text{in}} = R_x; R_{\text{out}} = R_z \tag{3.3}$$

For an ideal CCII, the input resistance (R_x) is zero and the output resistance (R_z) is infinite. Thus equation (3.3) can be expressed as below, for an ideal CCII.

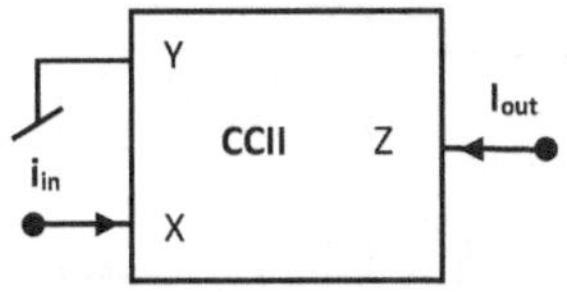

Figure 3.2 CCII as a current buffer.

$$i_{out} = i_{in}; R_{in} = 0; R_{out} = \infty \tag{3.4}$$

The CCII-based current buffer offers a zero input resistance and infinite output resistance. It provides isolation of signal source without loading, while driving the load circuit with no attenuation. The current buffer finds application in cascading sub-blocks with current input and current output, especially when impedance levels are not appropriate for such blocks. The loading problem is avoided by using current buffer (follower), while an accurate unity current transfer gain is ensured.

Practice Exercise 3.3 (solved). The CCII current buffer with 100 kΩ output resistance drives a load of 1 kΩ. If the input signal current is 100 μA, find the error in current delivered to load.

Solution: Let the output node voltage be V_o, then the input (i_{in}) flows through output resistance and load connected in parallel. The following current expression can be easily written:

$$\left(\frac{V_o}{100k} + \frac{V_o}{1k}\right) = 100\,\mu A$$

The current through load is $V_o/1k$; therefore the error can be found.

Practice Exercise 3.4: *Find the value of load for which the output current is within 2% error with the input signal.*

3.3 VOLTAGE TO CURRENT CONVERTER

Another interface circuit required as analog front end block is the voltage to current converter circuit. The CCII can be used for the purpose as shown in Figure 3.3. The desired features of this circuit are high input and high output impedances.

The circuit in Figure 3.3 shows an impedance Z_V at the X terminal. It is most often a resistor (R), used for conversion of input voltage signal to its

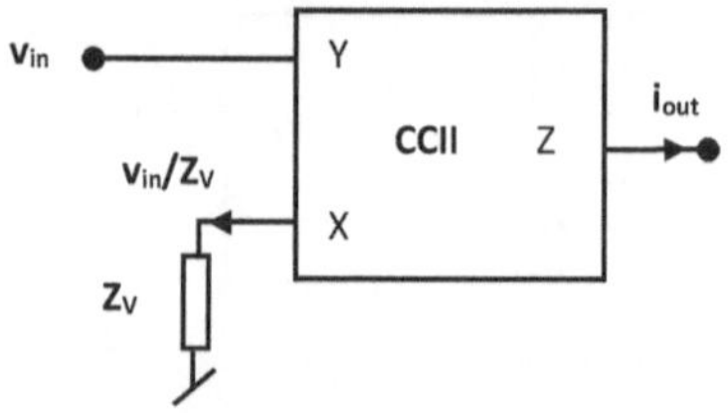

Figure 3.3 Voltage to current converter using a CCII.

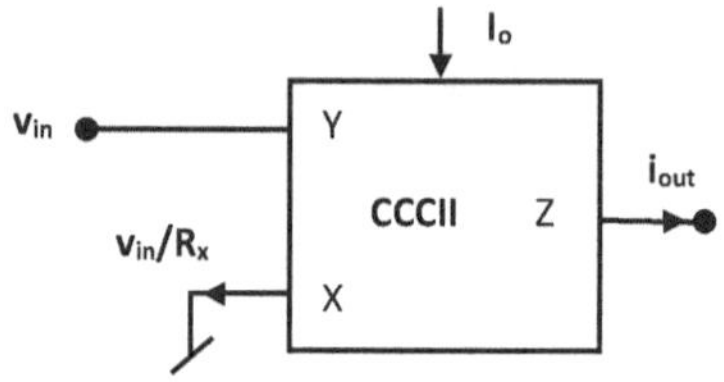

Figure 3.4 CCCII-based voltage to current converter.

current equivalent. The output current, input resistance and output resistance expression are given as equation (3.5) below.

$$i_{\text{out}} = v_{\text{in}}/R; R_{\text{in}} = \infty; R_{\text{out}} = \infty \tag{3.5}$$

The above discussed circuit is a positive polarity voltage to current converter employing a CCII+. A negative polarity circuit can be realized using a CCII–. The transconductance gain of the circuit is $1/R$. Another voltage to current converter circuit is presented in Figure 3.4. It is based on CCCII and provides an interesting electronically tuneable option.

The electronically tuneable circuit output $i_{out} = v_{in}/R_x$. The R_x is controlled by the CCCII bias current I_o. The bias current can be varied to change the effective value of transconductance $(1/R_x)$. The expression of R_x for a CMOS CCCII was introduced in Chapter 2. A positive polarity and negative polarity circuits are realized using a CCCII+ and a CCCII–, respectively. The electronically tuneable voltage to current converter finds useful applications in analog circuit design.

3.4 CURRENT TO VOLTAGE CONVERTER

The next interface circuit is current to voltage converter, which finds applications where the signal to be sensed is current while the load to be driven requires voltage signal. The CCII-based circuit for the purpose is shown in Figure 3.5, with the defining equation as given below.

$$v_{\text{out}} = -i_{\text{in}}/Z_V; R_{\text{in}} = R_x; R_{\text{out}} = Z_V \tag{3.6}$$

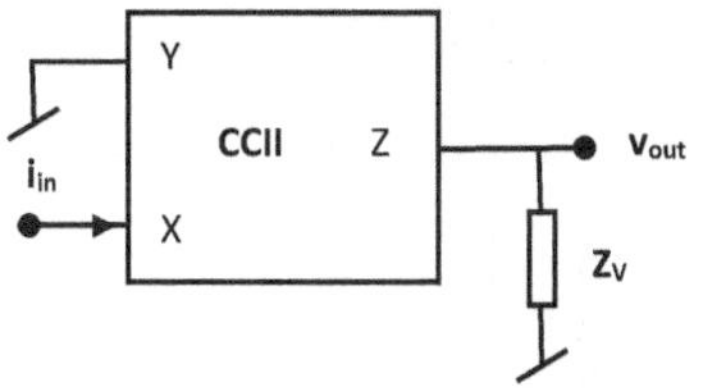

Figure 3.5 CCII as a current to voltage converter.

Equation (3.6) shows a negative sign in the output voltage expression. This is for CCII+-based circuit. For a CCII–, the output polarity is positive. The choice of Z_V in most applications is resistive. Thus for $Z_V = R$, the output voltage expression is $v_{out} = -i_{in}/R$ for the CCII+-based circuit. The input resistance is desirably low for the current input circuit shown in Figure 3.5. However, the output resistance is not desirably high. Instead, it depends on the load value (R). This circuit can be designed using an AD844 IC, introduced in Chapter 2, where the desirable low impedance voltage output is available at the W terminal of the AD844. The circuit has important applications in analog circuit design using current-mode approach.

Practice Exercise 3.5: *Show the connection diagram for an AD844-based current to voltage converter. Is the output resistance low? Justify.*

Practice Exercise 3.6: *Show the connection diagram for a CCII-based current to voltage converter. Realize the circuit using AD844 ICs. (Hint: Two AD844 are needed for a CCII–.)*

3.5 SINGLE ENDED VOLTAGE AMPLIFIERS

The current-mode active building blocks find useful applications in designing high performance amplifiers. This section covers single ended amplifiers based on various current conveyors. The voltage to current converter of Figure 3.3 can be a starting point to design an amplifier. The circuit when loaded by a resistor works as a simple single ended amplifier. The circuit is shown in Figure 3.6, with Z_1 and Z_2 to be specialized as resistors, namely R_1 and R_2, respectively.

The voltage transfer function for the circuit is as below.

$$\frac{v_{out}}{v_{in}} = \frac{R_2}{R_1} \tag{3.7}$$

Equation (3.7) can be used for designing the circuit for desired gain. A more realistic voltage gain expression is obtained by incorporating the parasitic model of CCII, discussed in Chapter 2. The inclusion of R_X and R_Z (for

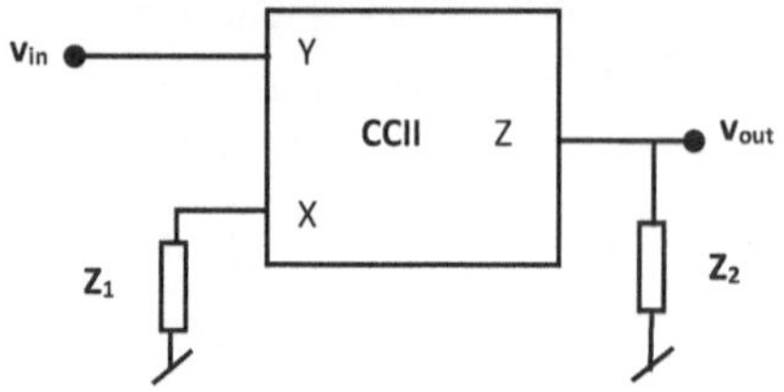

Figure 3.6 Voltage amplifier (for resistive impedances) using a CCII.

now ignoring parasitic capacitances) results in the following expression for voltage gain.

$$\frac{v_{\text{out}}}{v_{\text{in}}} = \frac{R_2 R_Z}{(R_2 + R_Z)(R_1 + R_X)} \tag{3.8}$$

Equation (3.8) suggests that the choice of external resistors must be such that the parasitic effects are minimized. This requires $R_2 < R_Z$, and $R_1 > R_X$.

The input resistance and the output resistance of the circuit are given as below.

$$R_{\text{in}} = R_Y; R_{\text{out}} = \frac{R_2 R_Z}{R_2 + R_Z} \tag{3.9}$$

Equation (3.9) suggests that the input resistance is high (ideally infinite), and the output resistance depends on the external resistance at Z terminal. It is to be noted that R_Y and R_Z are *Y*- and *Z*-terminal resistances, introduced in Chapter 2. For a choice of R_2 such that $R_2 << R_Z$, the output resistance is simply R_2. The circuit of Figure 3.6, when designed using an AD844 IC, can provide low output resistance at the buffered output (W) of the IC. The amplifier bandwidth expression is found by incorporating *Z*-terminal parasitic capacitance (C_Z). The *Y*-terminal parasitic capacitance does not affect the circuit because it is voltage-driven high input impedance terminal. The bandwidth expression is given below.

$$f_{-3\text{dB}} = \frac{1}{2\pi R_{\text{out}} C_Z} \tag{3.10}$$

The $f_{-3\text{dB}}$ as given in equation (3.10) is the frequency at which the low frequency gain (dc gain) drops by 3 dBs. It depends on the output resistance and *Z*-terminal parasitic capacitance. In order to design the amplifier with extended bandwidth, it is best desired to minimize the R_2 (hence R_{out}). The gain can be adjusted using R_1 without affecting the bandwidth.

Practice Exercise 3.7: *Design a CCII-based amplifier circuit for a voltage gain of 10 and bandwidth of 10 MHz. Assume $C_Y = 5$ pF, $C_Z = 7$ pF, $R_Z = 1$ MΩ and $R_X = 100$ Ω. (Hint: Use equation 3.10 to find R_{out} for given bandwidth, fix R_2 accordingly, then set R_1 for desired gain)*

Practice Exercise 3.8: *The CCII-based amplifier is designed for a gain of 5, but the actual gain obtained is 4.8. Comment on the reasons for this error.*

Practice Exercise 3.9: *If the $R_X = 100$ Ω, and the two available resistors are only of 5 and 10 kΩ, then what are the possible amplifier gains? Ignore other parasitic of CCII.*

The circuit of Figure 3.6 is assumed to be based on a CCII+, and realizes a non-inverting amplifier. If CCII– is used, then the realized transfer function is of inverting nature. Hence a CCII– can be used in place of CCII+ for designing an inverting amplifier. However, the non-inverting circuit of Figure 3.6 is compatible with a single AD844 IC. The inverting amplifier using a CCII– requires two AD844 ICs.

Practice Exercise 3.10: *Show the AD844-based inverting amplifier circuit and design it for a gain of –10.*

It is important for an amplifier circuit to provide variable gain which is electronically controlled. Thus, an electronically tuneable amplifier circuit can be designed using a CCCII instead of a CCII. The circuit is given in Figure 3.7. The voltage transfer function for the circuit is given as below.

$$\frac{v_{\text{out}}}{v_{\text{in}}} = \frac{R}{R_X} \tag{3.11}$$

In equation (3.11), R_X is intrinsic X terminal resistance controlled by I_o, which was introduced in Chapter 2. Therefore, the voltage gain can be tuned by varying the bias current of CCCII. The input and output resistance expressions and the bandwidth remain same as for the CCII-based circuit.

The electronically tuneable amplifier of Figure 3.7 finds interesting applications in analog signal processing, because variable gain amplifier (VGA) forms an important building block of typical information system.

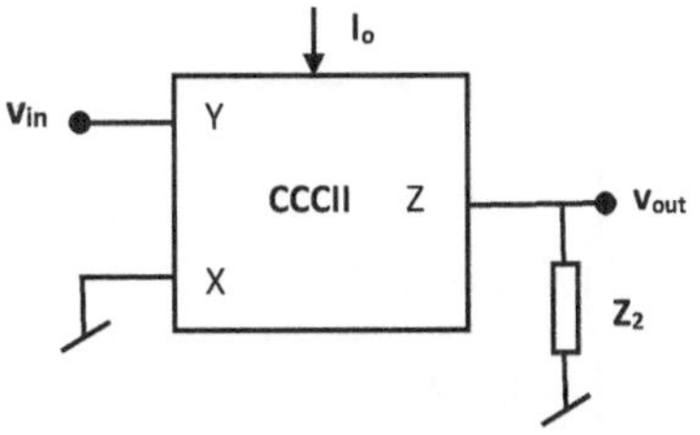

Figure 3.7 Electronically tuneable amplifier circuit for $Z_2 = R$.

3.6 SINGLE ENDED CURRENT AMPLIFIERS

The input and output signal of a current amplifier is current. It is characterized by the current transfer gain. An ideal current amplifier must possess low input and high output resistance. The CCII-based generalized current-mode circuit is shown in Figure 3.8.

It is the first circuit operating in current mode to be introduced in the book; therefore, it needs to be explained in detail. Although, the two impedances shown are general, the circuit finds many applications, amplifier being one of them. The analysis is carried using the CCII defining equation (please refer to Chapter 2). Since $i_y = 0$, the node voltage $v_y = i_{in}Z_2$. This voltage is conveyed to the X terminal; therefore $v_x = i_{in}Z_2$. Now the current through the X terminal is i_x, which is v_x/Z_1. The current through Z terminal (i_z) is i_{out}. Thus, the following relationship holds.

$$i_{out} = \frac{v_x}{Z_1} = \frac{i_{in}Z_2}{Z_1} \tag{3.12}$$

The current transfer function can be obtained from equation (3.12) as below.

$$\frac{i_{out}}{i_{in}} = \frac{Z_2}{Z_1} \tag{3.13}$$

In equation (3.13), if $Z_1 = R_1$ and $Z_2 = R_2$, the current transfer function reduces to the following.

$$\frac{i_{out}}{i_{in}} = \frac{R_2}{R_1} \tag{3.14}$$

The circuit realizes a current amplifier with gain (R_2/R_1), as given by equation (3.14). The CCII in Figure 3.8 is assumed to be of positive polarity (CCII+). The inverting amplifier is obtained with a CCII–. The input impedance of the circuit in Figure 3.8 is Z_2, while the output impedance is Z_z. In terms of amplifier input and output resistances, these values are R_2 and R_z, respectively. The input resistance is dependent on external resistor R_2, while

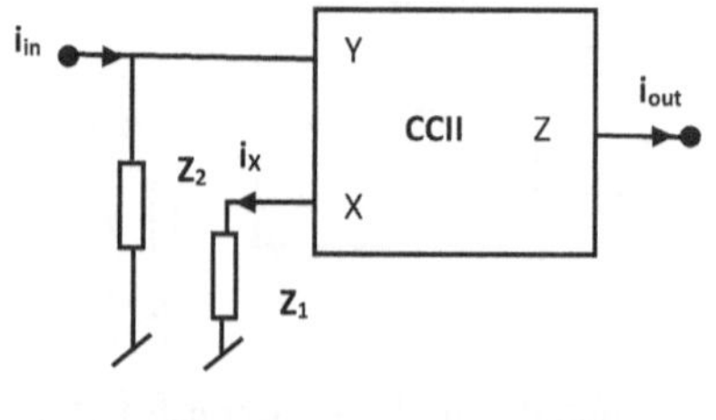

Figure 3.8 Current-mode amplifier circuit using a CCII.

the output resistance is R_z, which is high for a CCII. Thus, the circuit of Figure 3.8 provides the output at high impedance, which is desirable for connection with circuit operating in the current mode. The circuit using CCII+ is compatible with a single AD8844 IC, but the inverting topology requires two AD844 ICs.

Practice Exercise 3.11: *Show the inverting and non-inverting current-mode amplifier topologies using IC AD844. (Hint: Realize CCII using AD844, studied in Chapter 2)*

Practice Exercise 3.12: *Convert the circuit of Figure 3.8 into an electronically tuneable amplifier. (Hint: Use CCCII, with Z_1 replaced by a ground, R_x replaces R_1 for amplifier)*

The bandwidth calculation for the current amplifier can be done by considering the parasitic model of CCII. Z_1 and Z_2 are expressed as below for the purpose of bandwidth calculation.

$$Z_1 = R_1 + R_X; Z_2 = (R_2 // R_Y) // \frac{1}{C_Y} \tag{3.15}$$

From equation (3.15), the simplified expression of Z_2 is obtained as below.

$$Z_2 = \frac{R_2'}{sR_2'R_Y + 1}; R_2' = \frac{R_2 R_Y}{R_2 + R_Y} \tag{3.16}$$

Equation (3.16) can be used to find the location of poles, which decides the amplifier's bandwidth. The bandwidth expression is given as below.

$$f_{-3dB} = \frac{1}{2\pi R_2' C_Y} \tag{3.17}$$

Equation (3.17) suggests that the amplifier bandwidth depends on the Y-terminal parasitic impedance. The non-ideal dc current gain expression is also given as below.

$$\frac{i_{out}}{i_{in}} = \frac{R_2'}{R_1'}; R_1' = R_1 + R_X \tag{3.18}$$

Equation (3.18) is the non-ideal dc current gain, which incorporates the effect of parasitic resistances of CCII.

Practice Exercise 3.11: *The current amplifier circuit is designed using R_1 = 2 kΩ and R_2 = 4 kΩ. If the parasitic R_Y = 2 MΩ, C_Y = 5 pF, then find the current gain and bandwidth of the circuit.*

Practice Exercise 3.12: *Design the current amplifier for a bandwidth of 2 MHz and gain of 5, using parasitic values as in Exercise 3.11.*

3.7 INSTRUMENTATION AMPLIFIER

The need for processing differential signals with the advantage of common mode noise and interference rejection is fulfilled by dual-ended amplifiers. The instrumentation amplifier is one such circuit capable of amplifying differential signals. It finds useful applications in biomedical signal processing and instrumentation systems. It is useful in situations demanding greater noise and interference suppression. The differential signal processing has useful advantages in minimizing symmetry and mismatch problems in analog circuit layout during fabrication of integrated circuits. The instrumentation amplifier design using a current-mode building block is best possible using a differential voltage current conveyor (DVCC), already introduced in Chapter 2. A single DVCC is used to realize an instrumentation amplifier circuit. Most of the treatment in the chapter is based on the usage of single active building block for realizing various circuit functions. The instrumentation amplifier circuit using a DVCC is given in Figure 3.9, with the defining equation for DVCC being repeated here for easy reference.

$$i_{y1} = i_{y2} = 0, v_x = v_{y1} - v_{y2}, i_z = i_x \tag{3.19}$$

The circuit is analysed for the voltage transfer function using the defining equation (3.19). The same is given below.

$$\frac{v_o}{v_1 - v_2} = \frac{R_2}{R_1} \tag{3.20}$$

Defining the differential signal as $v_{id} = v_1 - v_2$, the equation (3.20) reduces to the following.

$$\frac{v_o}{v_{id}} = \frac{R_2}{R_1} \tag{3.21}$$

The above expression (3.21) is the differential mode gain. The common mode gain calculation requires non-ideal DVCC consideration. The main

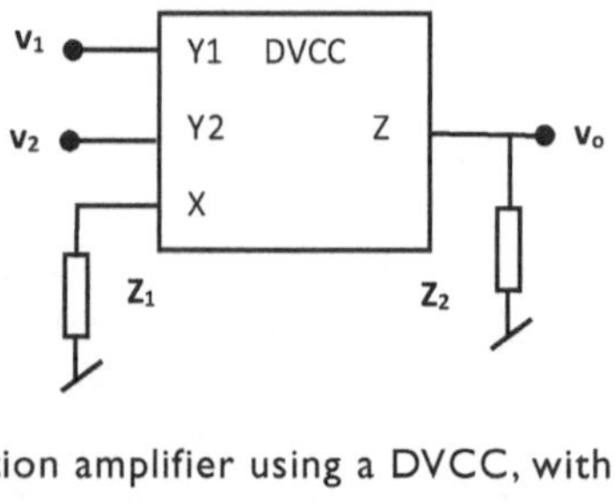

Figure 3.9 Instrumentation amplifier using a DVCC, with $Z_1 = R_1$ and $Z_2 = R_2$.

non-ideal parameters to be incorporated are finite voltage and current transfer gain of DVCC and the parasitic elements of the DVCC. These are not being considered for further elaboration. The ideal common mode gain is zero, thereby yielding and an infinite common mode rejection ratio (CMRR). It is important to mention that the CMRR is ratio of difference mode gain and common mode gain. It is an important figure of merit for an instrumentation amplifier. The instrumentation amplifier's bandwidth mainly depends on the Z-terminal parasitic capacitance, like the single ended amplifier discussed in earlier sections. For the sake of illustration, the DVCC-based circuit of Figure 3.9 is designed using 1 kΩ and 10 kΩ resistors R_1 and R_2, respectively. The input is applied in differential fashion at Y_1 and Y_2 terminals. The DVCC circuitry presented in Chapter 2 is simulated using 180 nm process parameters. The supply voltage used is ±1.5 V. The SPICE-based simulations are performed for transient response, where sinusoidal differential input is applied and the circuit output obtained with an expected gain of 10. The frequency of the applied signal is 1 MHz. The result is shown in Figure 3.10. It is clear that the differential input is $V_1 - V_2$, which is 5 mV, and the output is 50 mV. The peak values of sinusoidal signals are mentioned herein. The output spectrum is also shown plotted as Figure 3.11, which shows good selectively of signal frequency (1 MHz).

The instrumentation amplifier operating on current-mode signals is next presented using a single DVCC. The circuit is shown in Figure 3.12.

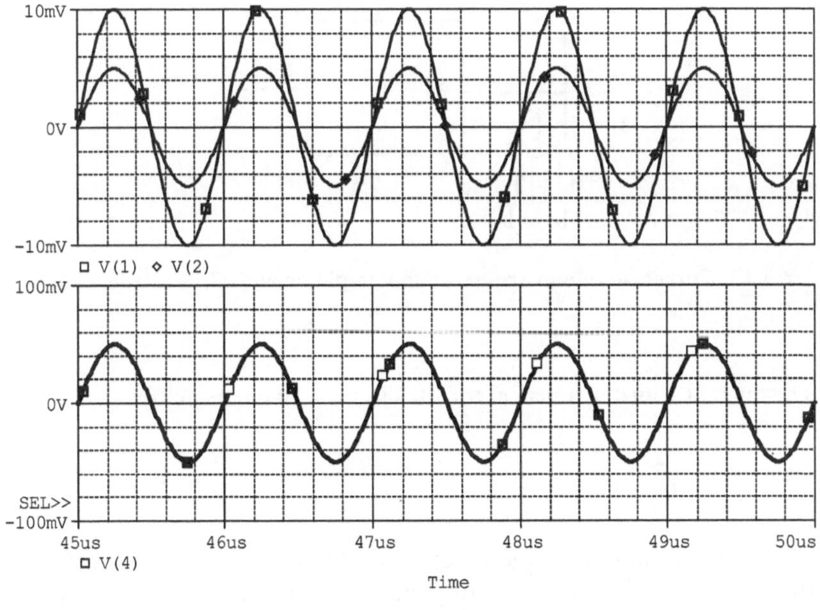

Figure 3.10 Inputs [$V(1)$ and $V(2)$] and output [$V(4)$] for designed circuit of Figure 3.9.

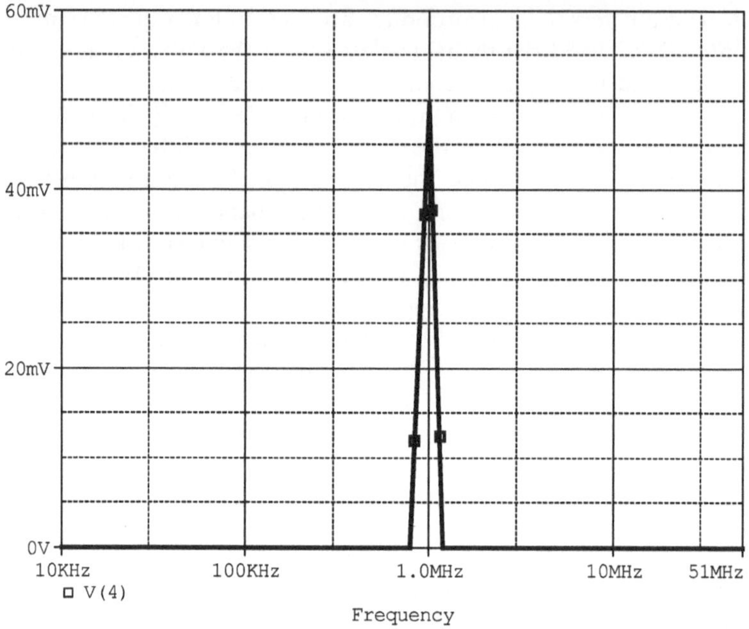

Figure 3.11 The output spectra showing good selectivity of signal frequency (1 MHz).

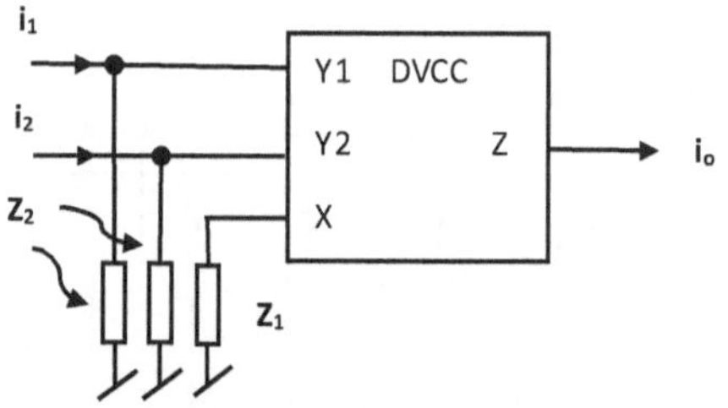

Figure 3.12 Current-mode Instrumentation amplifier with $Z_1 = R_1$ and $Z_2 = R_2$.

The current transfer function for the circuit is given as below.

$$\frac{i_o}{i_1 - i_2} = -\frac{R_2}{R_1} \tag{3.22}$$

The inverting nature of the function (3.22) is obtained using a DVCC with Z+ output. The non-inverting transfer function is realized using a DVCC with Z– output stage.

Practice Exercise 3.13: *The current-mode circuit of Figure 3.12 is designed using a DVCC with parasitic elements as $R_{Y1} = R_{Y2} = 1\ M\Omega$, $C_{Y1} = C_{Y2} = 5\ pF$. The external resistors used are $R_2 = 10\ k\Omega$ and $R_1 = 2\ k\Omega$. Calculate the differential current gain and amplifier bandwidth.*

Practice Exercise 3.14: *Derive the expression for current-mode instrumentation amplifier's bandwidth. (Hint: Use parasitic models of DVCC and analyse for its transfer function; the pole location gives the bandwidth)*

3.8 SUMMING AMPLIFIER

The summing of two analog signals is one of the important functions which need to be discussed using current-mode building blocks. The CCII-based summing amplifier circuit is shown in Figure 3.13. It is based on the current follower topology of Figure 3.2.

The output voltage expression is found as below.

$$v_{out} = -R_o\left(\frac{v_1}{R_1} + \frac{v_2}{R_2}\right) \tag{3.23}$$

For $R_1 = R_2 = R$, the above expression (3.23) reduces to the following.

$$v_{out} = -\frac{R_o}{R}(v_1 + v_2) \tag{3.24}$$

Equation (3.24) can be used to design the circuit. Both scaled summing or direct summing are possible with appropriate choice of resistors. The inverting nature of the function is obtained with a CCII+, while the non-inverting function can be obtained if CCII– is used. The circuit can also be extended for more than two inputs. However, the maximum current input at the X terminal would limit the number of inputs. The circuit can be easily realized using a single AD844 IC, with the advantage of buffered output at the W terminal. How to calculate the bandwidth expression for the circuit? This can be done by using parasitic model of DVCC. The output node parasitic, namely the Z-terminal parasitic capacitance would decide the amplifier bandwidth.

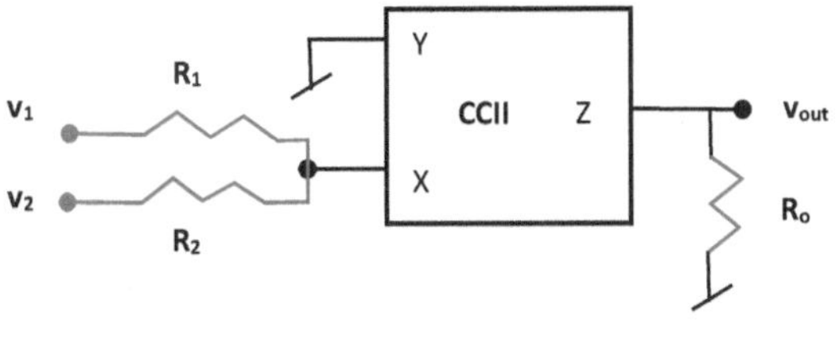

Figure 3.13 CCII as summing amplifier.

Practice Exercise 3.15: *Design the summing amplifier for* $v_{out} = -(1.2\ v_1 + 0.8v_2)$.

Practice Exercise 3.16: *For the above problem, if CCII has* $R_Z = 800\ k\Omega$, *then select* R_o *such that the output is not in error by more than 2%.*

3.9 FURTHER READING

The analog interface circuits and amplifiers circuits as discussed in previous sections are the basic building blocks of analog signal processing. For example, some of the basic building blocks were introduced at the time of current conveyor invention [5]. The circuits based on current control for performing simple electronic functions were reported in some of the first few works on this building block [9]. The basic functions of the differential voltage current conveyor were reported in another work [8]. The recent introduction of extra X current conveyor can be found in a relatively recent work [10]. The active only amplifiers are available in several different works by the author [11–13]. Interested readers are encouraged to explore further details on the specific topics from some of these works.

Chapter 4

Single time constant circuits

The first three chapters have laid the foundation of circuit design using current-mode building blocks. This chapter prepares the readers to understand single time constant circuits, which form the heart of the typical analog signal processing system. In this chapter, simple circuits in the form of lossy and lossless integrators are discussed, which find applications as frequency selective networks and form the basic building block for higher order circuits. The phase-shifting networks with all-pass filter nature are also covered. The circuits presented in the chapter include both voltage-mode and current-mode ones. The active building block used is CCII. The electronic tuning of the circuits is discussed and some examples of such circuits are further covered in detail. Thus, CCCII-based circuits are discussed for realizing the phase-shifting function. The IC compatibility of circuits is presented for readers to experimentally build the circuits as laboratory exercises.

4.1 LOSSY AND LOSSLESS INTEGRATORS

The basic analog circuits in the form of lossy and lossless integrators are introduced in this section. It is well known that such circuits are often built using conventional operational amplifier and passive components. The lossy and lossless integrator transfer functions are first given below for easy reference.

$$\frac{v_o}{v_{in}} = \frac{1}{1 + sT} \tag{4.1}$$

$$\frac{v_o}{v_{in}} = \frac{1}{sT} \tag{4.2}$$

Equation (4.1) is for the lossy integrator, whereas equation (4.2) is for the lossless integrator. Here T is the circuit time constant, which for an active-RC circuit is often $T = RC$. The circuits for the two functions can be realized

DOI: 10.1201/9781003403111-4

using CCII. Figure 4.1 shows the lossy and Figure 4.2 shows the lossless integrator circuit. The lossless integrator circuit uses one resistor and one capacitor, whereas a lossy integrator requires two resistors and a capacitor. The single capacitor requirement arises from the fact that a single time constant circuit realization is possible with a minimum of one capacitor.

The voltage transfer function for the lossy integrator circuit is given below.

$$\frac{v_o}{v_{\text{in}}} = \frac{\left(R_2/R_1\right)}{1 + sR_2C} \tag{4.3}$$

Equation (4.3) needs some explanation. The dc gain of the circuit is (R_2/R_1). The frequency at which the gain drops to 0.707 of the dc gain is given by $\omega_p = 1/R_2C$. This is also called the pole frequency. The gain drops at a rate of –20dB/decade. The lossy integrator provides a finite gain at low frequency, unlike a lossless integrator. The lossy integrator also works as a low-pass filter of the first order. The nature of transfer function is of non-inverting type with CCII+. The inverting function can be realized using a CCII. The lossless integrator of Figure 4.2 is characterized by the following transfer function.

$$\frac{v_o}{v_{\text{in}}} = \frac{1}{sRC} \tag{4.4}$$

The dc gain for the lossless function of equation (4.4) is infinite. It is easy to verify the same by substituting $s = j\omega$, and putting $f = 0$. The integrator

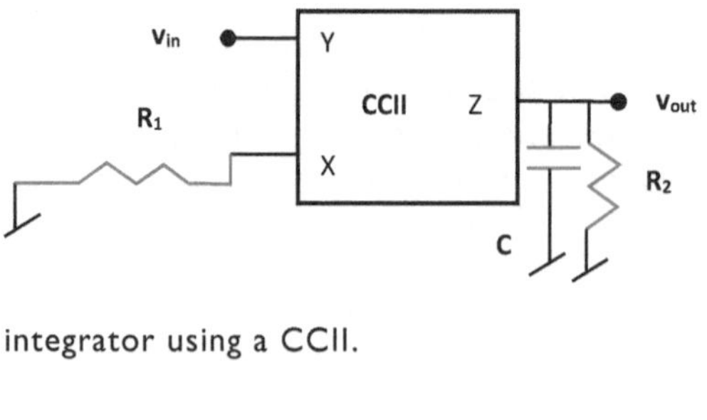

Figure 4.1 Lossy integrator using a CCII.

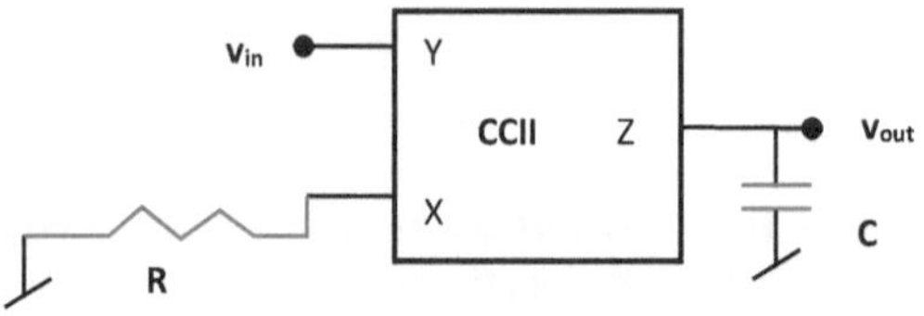

Figure 4.2 Lossless integrator using a CCII.

frequency is defined as the frequency at which gain becomes unity. It is denoted and expressed as: $\omega_{int} = 1/RC$. The gain decreases at –20dB/decade from low frequency value to high frequencies. The inverting nature of transfer function can be obtained if CCII– is employed in place of a CCII+. Both the functions of Figures 4.1 and 4.2 can be easily breadboarded using a single IC AD844. The circuits provide high input impedance in both cases. The IC-based realizations would also benefit from low output resistance, as the output would be available at the *W* terminal of AD844. The inverting functions using the same topologies require two ICs in each case, hence may not be an economical solution. There are other possible circuit topologies for inverting functions using a single IC.

Examples of such circuits performing lossy and lossless integrator function with inverting transfer function are shown in Figures 4.3 and 4.4, respectively. Neither of the two circuits possesses high input resistance, because the input signal is connected through a resistor in each case. The circuits are based on CCII+.

The inverting lossy integrator transfer function is given below as equation (4.5).

$$\frac{v_o}{v_{in}} = -\frac{\left(R_2/R_1\right)}{1 + sR_2C} \tag{4.5}$$

The inverting lossless integrator transfer function is given below as equation (4.6).

$$\frac{v_o}{v_{in}} = -\frac{1}{sRC} \tag{4.6}$$

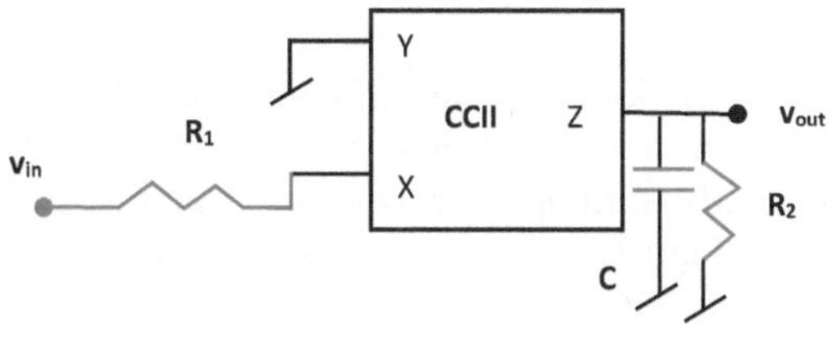

Figure 4.3 Inverting lossy integrator circuit using a CCII.

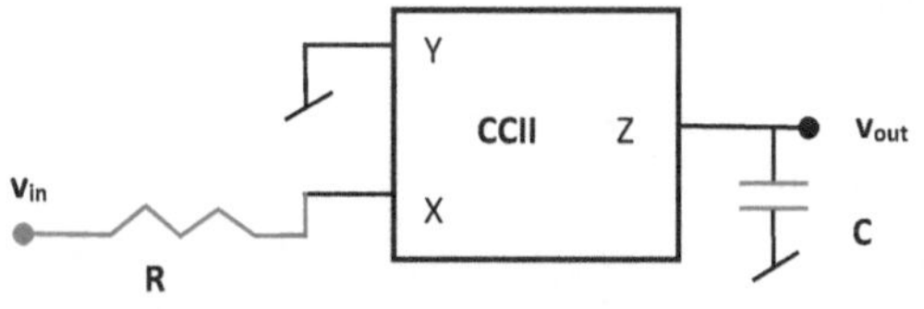

Figure 4.4 Inverting lossless integrator using a CCII.

The lossy integrator works as a low-pass filter allowing lower frequency signals to pass with large gain, while attenuating higher frequency signals. The pole frequency of the circuit is given as $\omega_p = 1/R_2C$. The dc gain of the lossy integrator depends on the ratio of two resistors. The dc gain of the lossless integrator is infinite. Both the circuits with inverting nature of transfer function can be realized using a single AD844 IC, with the additional advantage of buffered output.

Practice Exercise 4.1: *Show the AD844-based realizations for the four integrator circuits presented in the section (4.1).*

Practice Exercise 4.2: *Design the two lossless integrator circuits (Figures 4.1 and 4.4) for unity gain frequency of 1 MHz in each case. (Hint: Select C and find R)*

Practice Exercise 4.3: *Design the lossy integrator circuits (Figures 4.2 and 4.3) for dc gain as 10, and pole frequency as 1 MHz in each case.*

4.2 LOSSY AND LOSSLESS DIFFERENTIATORS

The next single time constant circuit to be considered is the differentiator. Two types of differentiator circuits are available, namely, lossy and lossless differentiators. The generalized transfer functions for the two functions are given as below.

$$\frac{v_o}{v_{in}} = \frac{sT}{1+sT} \tag{4.7}$$

$$\frac{v_o}{v_{in}} = sT \tag{4.8}$$

Both the transfer functions as given in equations (4.7) and (4.8) can be realized using a CCII in each case. It may be noted that $T = RC$ for an active-RC network. The CCII-based non-inverting lossy and lossless differentiator circuits are shown in Figures 4.5 and 4.6, respectively.

The transfer function for the lossy differentiator circuit is given below.

$$\frac{v_o}{v_{in}} = \frac{sRC_1}{1+sRC_2} \tag{4.9}$$

For equal valued capacitors ($C_1 = C_2 = C$), the above equation (4.9) reduces to the following.

$$\frac{v_o}{v_{in}} = \frac{sRC}{1+sRC} \tag{4.10}$$

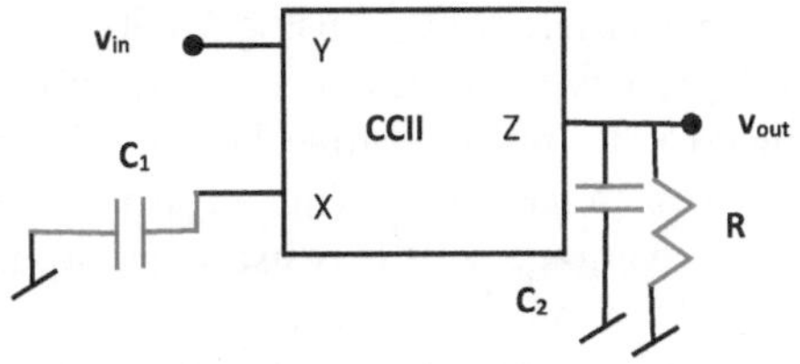

Figure 4.5 Lossy differentiator using a CCII.

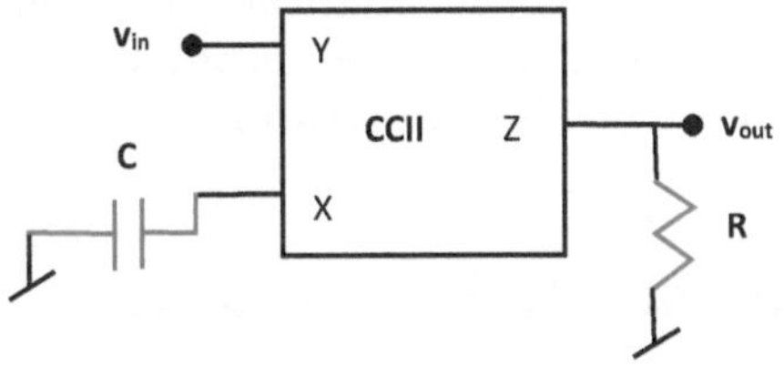

Figure 4.6 Lossless differentiator using a CCII.

The circuit works as a high-pass filter, allowing higher frequency to pass with large gain, while attenuating the lower frequency signals. This is evident from the transfer function of equation (4.10). The pole frequency of the circuit is given as $\omega_p = 1/RC$. The lossless differentiator circuit of Figure 4.6 exhibits a zero at dc and the gain increases at higher frequency. The transfer function is expressed as below.

$$\frac{v_o}{v_{in}} = sRC \tag{4.11}$$

The time constant of the circuit with transfer function given in equation (4.11) is as $T = RC$. The two circuits of lossy and lossless differentiator have the advantage of high input resistance, and once realized using AD844 IC, they also exhibit low output resistance, since the output can be tapped from the W terminal of the AD844 IC.

Practice Exercise 4.4: *Derive the expression for voltage transfer function for Figure 4.5, if resistor is open circuited.*

Practice Exercise 4.5: *Design the lossy differentiator for a pole frequency of 2 MHz. Assume a capacitor of 100 pF to be available.*

Practice Exercise 4.6: *For the lossless differentiator with C = 100 pF, and R = 2 kΩ, plot the gain response from 10 Hz to 10 MHz, taking data at a decade interval. (Hint: replace s = jω, and substitute the value of frequency, say, 10, 100, 1 k, 10 kHz etc.)*

The design of inverting differentiators using CCII+ is possible, similar to the way inverting integrators were designed in the previous section. The readers are encouraged to attempt the same. The circuits so designed would not exhibit high input resistance. Their realization using a single AD844 is also possible, with the advantage of low resistance output. The readers are advised to work out this as practice exercises.

The single time constant circuits discussed in the section cover the useful analog building blocks like integrators and differentiators. The lossy networks in the two categories also find useful filtering applications as low-pass and high-pass networks of first order. The section to follow extends the discussion to another important filtering network, which also falls under single time constant circuits. It can be viewed as a combination of low-pass and high-pass networks with different polarities.

4.3 ACTIVE PHASE-SHIFTING NETWORKS

The next class of circuits to be presented fall under the category of phase-shifting networks. It is well known to the readers that capacitors and inductors are used in conjunction with resistors for realizing such networks. The phase shifters are broadly of two types, namely lead and lag networks. The applications of such networks are diverse, ranging from the design of analog sub-systems, instrumentation and communication circuits. Confining to the design of analog sub-systems, the use of operational amplifier along with passive components is well known to the readers. Such circuits can be further defined in terms of their characteristics: namely a constant gain at all frequency, while offering a phase shift between the output and the input, which varies with the frequency of signal. Such circuits are also referred to as all-pass filters. The simplest of such circuits can be of first order, which requires only a single capacitor and resistor(s). The limitations of operational amplifier-based phase shifters are attributed to the inherent voltage-mode operation, gain-bandwidth product and high frequency restrictions. The use of current-mode techniques for realizing phase-shifting circuits offers advantages of lower voltage operation, higher achievable bandwidth and accuracy of obtained phase-shift form such networks. The CCII-based first such circuit is shown in Figure 4.7. It employs two resistors and a capacitor, besides a single CCII+. The circuit is based on the current follower topology, with grounded Y terminal. The first order all-pass filter circuit of Figure 4.7 is characterized by the following voltage transfer function [14].

$$\frac{v_o}{v_{in}} = \frac{s - \left(\frac{1}{R_1 C} - \frac{1}{R_2 C}\right)}{s + \frac{1}{R_2 C}} \tag{4.12}$$

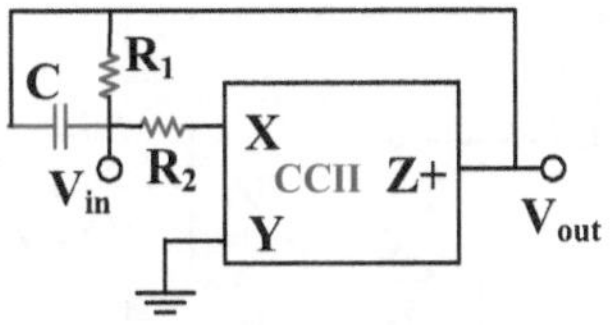

Figure 4.7 CCII+-based first order all-pass filter circuit [14].

For $R_1 = R_2/2 = R/2$, the above equation (4.12) reduces to the following.

$$\frac{v_{\text{out}}}{v_{\text{in}}} = \frac{s - \dfrac{1}{RC}}{s + \dfrac{1}{RC}} \tag{4.13}$$

Equation (4.13) represents a standard first order all-pass filter transfer function. The gain magnitude and the phase shift between the output and the input can be expressed as below.

$$\left|\frac{v_{\text{out}}}{v_{\text{in}}}\right| = 1; \varnothing = 180^{\circ} - 2\tan^{-1}(\omega RC) \tag{4.14}$$

The circuit provides a constant unity gain at all frequencies, while providing a frequency dependent phase shift, which varies from 180 to 0 degree. The low frequency signals experience a phase shift of 180 degrees, while as the frequency is increased, the phase shift decreases. The pole frequency is defined as the frequency where the phase shift becomes 90°. This occurs for $\omega = 1/RC$. At this value, equation (4.14) suggests a phase shift of 90°. A further increase in signal frequency results in the output signal attaining the phase of the input, implying a zero-degree phase shift. It may be noted that $\omega = 2\pi f = 1/RC$, where, 'f' is the frequency of signal in Hertz (Hz).

Practice Exercise 4.7: *Design the phase shifter of Figure 4.7 for providing a phase shift of 90° to 0.5 MHz signal. Hint: Assume C = 100 pF, and use ω = 1/RC to find R, hence R_1 and R_2.*

The next circuit for performing phase-shifting operation is based on CCII– and is shown in Figure 4.8 [15]. The circuit requires two resistors and a capacitor and is described by the following voltage transfer function.

$$\frac{v_{\text{out}}}{v_{\text{in}}} = \frac{s - \dfrac{1}{R_1 C}}{s + \dfrac{1}{R_2 C}} \tag{4.15}$$

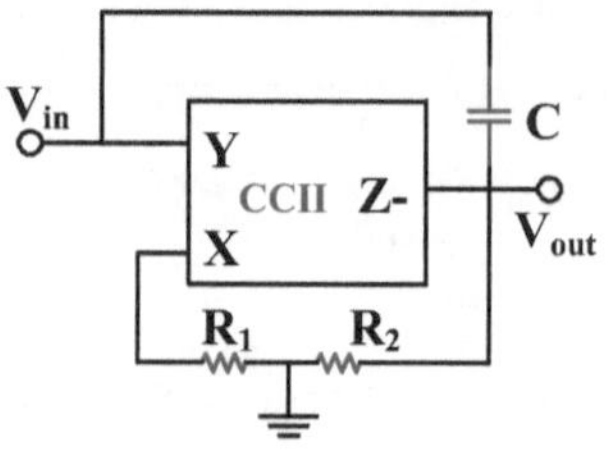

Figure 4.8 CCII−-based first order all-pass filter circuit [15].

The above transfer function of equation (4.15) can be simplified by assuming $R_1 = R_2 = R$. The resulting transfer function then reduces to equation (4.16) as below.

$$\frac{v_{\text{out}}}{v_{\text{in}}} = \frac{s - \dfrac{1}{RC}}{s + \dfrac{1}{RC}} \tag{4.16}$$

Therefore, the circuit also realizes a voltage transfer function with phase-shifting characteristics, while preserving the gain magnitude. The gain magnitude and the phase shift between the output and input signal are same as given by equation 4.14.

Practice Exercise 4.8: *For the circuit shown in Figure 4.9 analyse the voltage transfer function and find the gain magnitude and phase-shift expressions.*

The next circuit for active phase shifter to be discussed is for current input and current output. Hence it can be referred to as current-mode first order all-pass filter. It is designed using a CCII− with two $Z-$ stages, a resistor and a capacitor. The circuit is shown in Figure 4.10. The circuit can be analysed using the CCII− description, for the following current-mode transfer function given as equation (4.17).

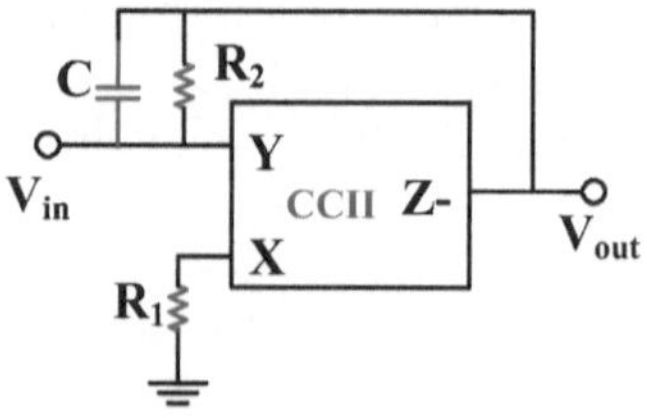

Figure 4.9 CCII−-based all-pass filter circuit [16].

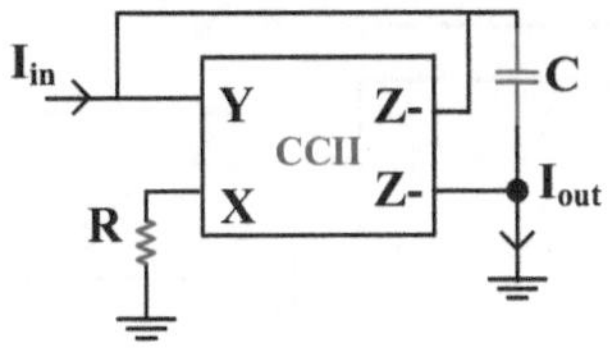

Figure 4.10 Current-mode all-pass filter circuit.

$$\frac{I_{out}}{I_{in}} = \frac{s - \frac{1}{RC}}{s + \frac{1}{RC}} \tag{4.17}$$

The output node is referenced to ground, meaning that the current needs to be measured using extra sensing element. This aspect needs more elaboration. The shorted second *Z*– stage and capacitor termination (Figure 4.10) are shown connected to ground. Moreover, the output current is also tapped from this shorted node. From a practical perspective, it is not easy to tap the output without an additional sensing element. Therefore, an additional current conveyor is required as a current follower (Chapter 2) to actually measure the output current. As far as the gain magnitude and phase-shift expressions are concerned, these are same as in the two preceding voltage-mode circuits. The only difference is that the input and output signals are current. The advantage of this circuit as compared to the preceding two circuits lies in using a single resistor. The earlier presented voltage-mode circuits require two resistors, with certain matching. This circuit is the active-RC equivalent of the active-C version available in ref. [17].

Practice Exercise 4.9: *Derive the current transfer function for the current-mode circuit of Figure 4.10.*

Practice Exercise 4.10: *Design the circuit of Figure 4.10 for a pole frequency of 1 MHz, using a capacitor of 100 pF.*

The last circuit to be discussed under this section is another current-mode circuit with two outputs, providing a phase shift of 180 to 0 and 0 to –180 degrees, respectively. The is based on CCII+ and uses a single resistor and capacitor. The second CCII with dual outputs (current follower) is used for sensing the output currents at high impedance levels. The circuit is shown in Figure 4.11. The circuit analysis yields the two current-mode transfer functions given in equations 4.18 and 4.19. The electronically tuneable version of this circuit using CCCIIs is available in ref. [18]; however, the circuit of Figure 4.11 is based on employing CCIIs. The circuit is therefore an

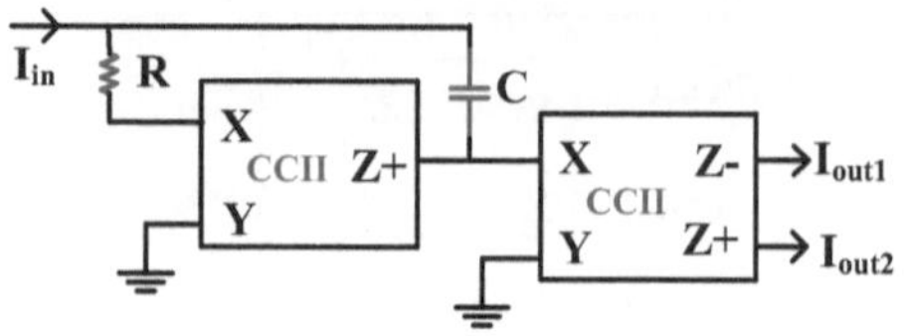

Figure 4.11 Another current-mode phase-shifting circuit with two outputs.

active-RC phase-shifting network. The two current transfer functions (4.18 and 4.19) are given as below.

$$\frac{I_{\text{out1}}}{I_{\text{in}}} = \frac{s - \frac{1}{RC}}{s + \frac{1}{RC}} \tag{4.18}$$

$$\frac{I_{\text{out2}}}{I_{\text{in}}} = -\frac{s - \frac{1}{RC}}{s + \frac{1}{RC}} \tag{4.19}$$

The transfer function ($I_{\text{out2}}/I_{\text{in}}$) is different from the previously discussed function by an inverting sign, which actually changes the phase characteristics of the circuit. The gain magnitude is same for both the outputs, but the phase-shift expression for the second output (I_{out2}) is as given below.

$$\varnothing = -2\tan^{-1}(\omega RC) \tag{4.20}$$

Equation (4.20) suggests that the phase shift between the output and the input signal is zero at low frequency, while it approaches to –180 degree at very high frequency. The frequency at which the phase shift is –90° is known as the pole frequency of the all-pass filter. It is expressed as $\omega = 1/RC$. The benefit of the circuit of Figure 4.11 lies in the fact that both types of phase shifts (positive and negative) are obtained. In other words, the two different outputs can provide leading and lagging phase shifts with respect to the input signal simultaneously.

In all the above discussed circuits for realizing phase-shifting function, the resistor and capacitor network is used alongside the current conveyors. The active-RC circuits can be designed for given specifications, namely the desired phase shift by varying the passive component values. Some of the practice exercises also confirm this assertion. The design is performed by assuming a value of capacitor and varying the resistor(s) for obtaining the desired specifications. The class of circuits where the resistors are replaced by active-equivalents are desirable from electronic tuning perspective, while also reduces the required chip area and noise by eliminating resistors. Such circuits are discussed in the following section.

4.4 ELECTRONICALLY TUNEABLE CIRCUITS

The active-C networks for realizing the phase shifter using current-mode techniques are considered herein. The first circuit for the purpose is shown in Figure 4.12. It requires a single capacitor and two CCCIIs. The core of the circuit is the active-RC first order all-pass filter of Figure 4.9 [16]. The resistor at the X terminal (R_1) is simply replaced by the grounded X terminal in the CCCII counterpart. The floating resistor (R_2) is replaced by the active equivalent as CCCII–(2). The implementation of a floating resistor using CCCII– is separately shown in Figure 4.13. The resulting circuit (Figure 4.12) becomes a tuneable all-pass filter of first order.

First, the floating tuneable active resistor of Figure 4.13 is considered. The analysis of the circuit is performed using the marked input voltage and current (V_i and I_i), respectively. The X and Z– terminal currents are given in equation (4.21) below.

$$I_x = \frac{V_1 - V_2}{R_x}; I_z = I_i \tag{4.21}$$

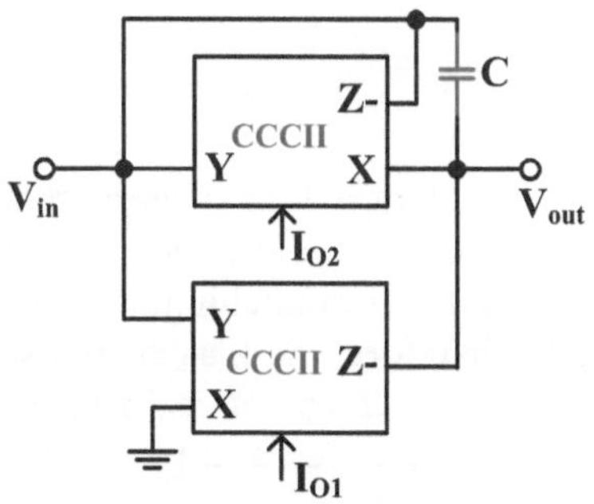

Figure 4.12 Electronically tuneable phase-shifter circuit [16].

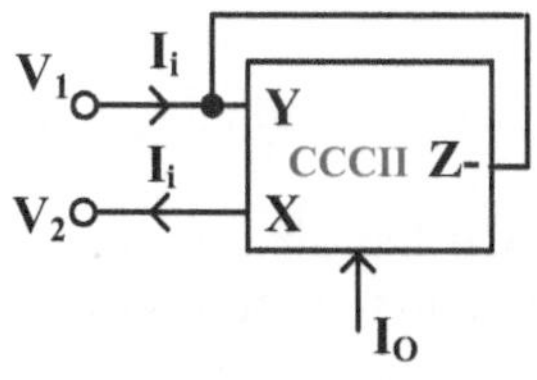

Figure 4.13 Electronically tuneable floating active resistor.

Remembering the current relationship for CCCII–, namely $i_z = -i_x$, and noticing the direction of current entering into $Z-$ terminal, while leaving the X terminal, we get the following expression.

$$\frac{V_i - V_2}{I_i} = R_x \tag{4.22}$$

Equation (4.22) represents the input resistance of the circuit in Figure 4.13. Thus, the circuit of Figure 4.13 realizes a tuneable floating active resistor, whose value can be tuned by the bias current of CCCIIs (I_o). Now coming to the phase-shifter circuit of Figure 4.12, the voltage transfer function is given as below.

$$\frac{v_{\text{out}}}{v_{in}} = \frac{s - \left(\dfrac{1}{R_{x2}C} - \dfrac{1}{R_{x1}C}\right)}{s + \dfrac{1}{R_{x1}C}} \tag{4.23}$$

Assuming $R_{x2} = R_{x1}/2$ in equation (4.23), we get the simplified transfer function as below.

$$\frac{v_{\text{out}}}{v_{\text{in}}} = \frac{s - \dfrac{1}{R_{x1}C}}{s + \dfrac{1}{R_{x1}C}} \tag{4.24}$$

The transfer function in equation 4.24 suggests that the circuit works as a phase shifter with electronic tuning of the pole frequency through the bias current of CCCIIs. The condition of matching for obtaining the above function demands $I_{o1} = I_{o2}/2$. Therefore, the bias currents need to be appropriately set for fulfilling the requirement of realizing the first order all-pass transfer function. The gain magnitude and phase-shift expressions for the circuit are given in equation (4.25).

$$\left|\frac{v_{\text{out}}}{v_{\text{in}}}\right| = 1; \varnothing = 180^{\circ} - 2\tan^{-1}\left(\omega R_{x1}C\right) \tag{4.25}$$

The circuit tuning aspect can be studied by expressing the R_{x1} in terms of the bias current, namely, $R_{x1} = V_T/2I_{o1}$. The readers are advised to refer to the earlier chapters, where CCCII has already been described. The phase-shift expression can therefore be written in terms of the bias current as below.

$$\varnothing = 180^{\circ} - 2\tan^{-1}\left(\omega C V_T / 2I_{o1}\right) \tag{4.26}$$

Equation (4.26) justifies the electronic tuning ability of the phase-shifting circuits, where the bias current I_{o1} can be varied to control the phase shift.

Practice Exercise 4.11: *Design the phase-shifting circuit of Figure 4.12 for providing a varying phase shift from 60° to 120° to the input sinusoidal signal at 0.5 MHz frequency. (Hint: Select a fixed capacitor value (say 100 pF); use equation (4.26) to find I_{o1} (hence I_{o2}) for desired phase shifts, at interval of 20°)*

The last circuit to be described under the category of electronically tuneable phase shifters is easily derived from the earlier discussed circuit of Figure 4.11. It is a current-mode circuit with two outputs providing different phase shifts. The circuit is shown in Figure 4.14. It is the tuneable version of the active-RC circuit of Figure 4.11. The CCII therein has been replaced by a CCCII, with resistor at the X terminal. The second current conveyor acts as a current follower with two phase-inverted outputs. This allows the output availability at high impedance nodes, which is ideal for cascading in current-mode operation.

The circuit is described by the following two current transfer functions as given in equations (4.27 and 4.28)

$$\frac{I_{out1}}{I_{in}} = \frac{s - \dfrac{1}{R_x C}}{s + \dfrac{1}{R_x C}} \tag{4.27}$$

$$\frac{I_{out2}}{I_{in}} = -\frac{s - \dfrac{1}{R_X C}}{s + \dfrac{1}{R_x C}} \tag{4.28}$$

The phase-shift expressions for the two outputs are given as equations (4.29) and (4.30).

$$\varnothing 1 = 180^{\circ} - 2\tan^{-1}(\omega R_x C); \tag{4.29}$$

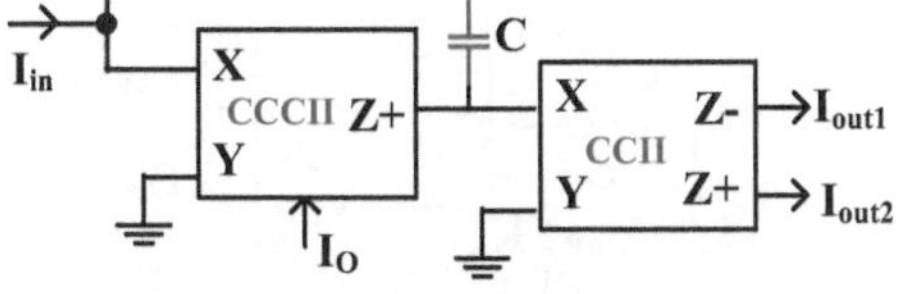

Figure 4.14 Electronically tuneable phase-shifter circuit with two outputs [18].

$$\varnothing 2 = -2\tan^{-1}(\omega R_x C); \quad (4.30)$$

The pole frequency and the phase shift can be electronically controlled by varying the bias current of the CCCIIs. The readers are reminded of the pole-frequency expression, which in this case is $\omega_p = 1/R_x C$. For instance, if the capacitor is chosen as 100 pF, and given $V_T = 25$ mV, the circuit can be designed for desired phase shifts. As an example, if the phase shift of 90° and –90° is desired at outputs I_{out1} and I_{out2}, for a signal of 1 MHz, the bias current can be calculated using the pole-frequency expression. The value of R_x is found as 1.59 KΩ, which yields the bias current value ($R_x = V_T/2I_o$) as 7.85 μA. The readers are advised to repeat this exercise for the capacitor value of 50 pF.

4.5 FIRST ORDER MULTIFUNCTION FILTER

Most of the presented circuits in the preceding sections are single input, single output type first order circuits. Circuits with multiple outputs are of interest since the same topology can be used for performing distinct functions simultaneously, as per the desired application. Such filters are called multifunction filters. Although the term 'universal filter' is also used in conjunction with first order sections for such filters. The first order universal filter is one which provides the low-pass, high-pass and all-pass responses simultaneously. One such circuit based on DVCC is shown in Figure 4.15 [19]. It requires two DVCCs and passive components to realize all three filter functions in voltage mode. Simultaneously, it provides current outputs for all-pass functions.

The filter transfer functions are given below.

$$\frac{V_{AP}}{V_{in}} = -\frac{s - \frac{1}{RC}}{s + \frac{1}{RC}}; \frac{V_{LP}}{V_{in}} = \frac{\frac{1}{RC}}{s + \frac{1}{RC}}; \frac{V_{HP}}{V_{in}} = \frac{s}{s + \frac{1}{RC}} \quad (4.31)$$

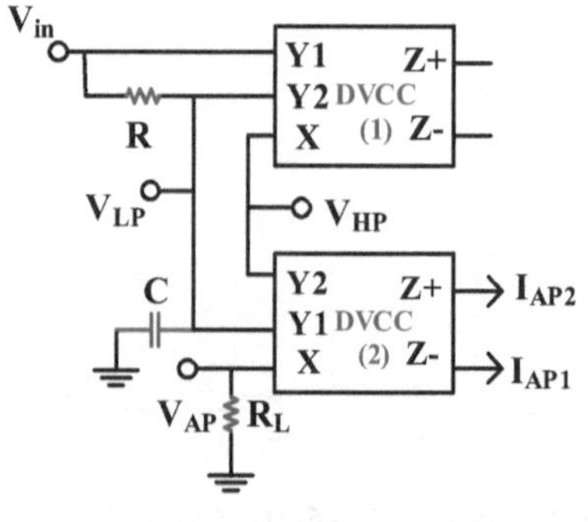

Figure 4.15 DVCC-based multifunction first order filter [19].

$$\frac{I_{AP1}}{V_{in}} = \left[\frac{1}{R_L}\right]\frac{s-\frac{1}{RC}}{s+\frac{1}{RC}}; \frac{I_{AP1}}{V_{in}} = -\left[\frac{1}{R_L}\right]\frac{s-\frac{1}{RC}}{s+\frac{1}{RC}} \tag{4.32}$$

Equation (4.31) represents three first order filters with voltage transfer functions, whereas equation (4.32) is the all-pass filter functions with voltage input and current outputs. Thus, a total of five filter responses can be obtained simultaneously from the multifunction filter circuit. The Z stages of the DVCC(1) are unused in the circuit. Another circuit with complete usage of DVCC stages, and providing all three current outputs, besides the three voltage outputs is next studied. The circuit is shown in Figure 4.16, and provides six outputs: three voltages and three currents simultaneously. Therefore, the multifunction filter may be called a single input six outputs first order filter. The three voltage outputs are similar to the previous circuit; hence equation (4.31) is valid for the circuit of Figure 4.16. The additional three filter functions are given as below.

$$\frac{I_{AP}}{V_{in}} = \left[\frac{1}{R_L}\right]\frac{s-\frac{1}{RC}}{s+\frac{1}{RC}}; \frac{I_{LP}}{V_{in}} = \left[\frac{1}{R_L}\right]\frac{\frac{1}{RC}}{s+\frac{1}{RC}}; \frac{I_{HP}}{V_{in}} = -\left[\frac{1}{R_L}\right]\frac{s}{s+\frac{1}{RC}} \tag{4.33}$$

Therefore, the circuit can be referred to as a complete first order section with voltage input and both voltage as well as current outputs. This is clear from the transfer functions of equation (4.33). The DVCC circuitry in CMOS was discussed in earlier chapters. The resistor(s) in the circuit can be replaced by MOSFET-based active resistors. The capacitor in the circuit is in grounded form. The designed capacitor value may be chosen to be in the feasibility range, say in tens of picofarads only. With these considerations, the circuit is ideal for integration in CMOS technology. The availability of such multifunction filters can add to the flexibility in design of larger analog

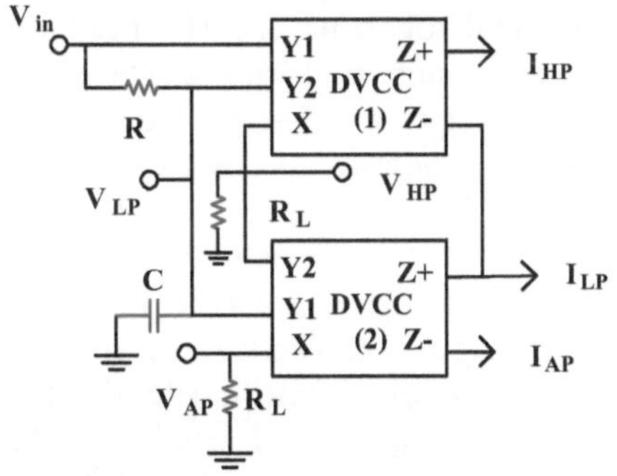

Figure 4.16 Single input six outputs multifunction first order filter circuit [19].

circuits, which utilize the first order sections as basic blocks. The advantage gained by replacing resistor(s) with their MOS equivalents serves another purpose as well. The active resistors are tuneable through external control voltage. Therefore, the circuit can be made electronically tuneable. The adoption of such a circuit by chip designers could be a realistic future problem, best left for industry professionals. It would be interesting to know the availability of such chips in the coming future. For the sake of engineering graduates, the interesting problem of designing the circuit using available CAD tools is a natural exercise to be undertaken. The design idea may begin by assuming a certain feasible capacitor value and meeting the desired specification, in terms of pole frequency. The circuit layout with the upcoming/available technology nodes in use may be another Master's level exercise, which can be undertaken as project work. Another possible laboratory exercise is to be mentioned. This is the circuit assembly using off-the-shelf components. It is well known at this stage of the text that a DVCC can be realized using IC AD844. The circuit can be bread-boarded by enthusiast readers to verify its functionality in laboratory conditions.

4.6 IC REALIZATIONS OF CIRCUITS

The single time constant circuits covered in the preceding sections are based on CCIIs or CCCIIs. It has already been studied in the previous chapters that a CCII+ can be realized using a single AD844 IC, whereas a CCII− is realized using two AD844 ICs. As far as CCCIIs are concerned, no equivalent IC realization is yet shown in this book, and shall not be attempted. Therefore, all the described circuits based on CCIIs can be easily built using AD844 ICs. The integrators, differentiators and all-pass filters, operating in voltage mode or current mode, can therefore be experimentally tested in laboratory environment. This provides useful opportunity to the students at various levels to actually bread-board the circuits given in the chapter. For example, the all-pass filter circuit of Figure 4.9 requires two AD844 ICs, two resistors and one capacitor for testing. The circuit design begins with the desired specifications, say a pole frequency of 400 kHz. The first step in design is the choice of capacitor value, say 100 pF. The value of two resistors can then be calculated using the formula. The circuit function realization requires $R_2 = R_1/2$ (refer to practice problem 4.8). The values of two resistors are found from the pole-frequency expression ($\omega_p = 1/R_1C$). The value obtained for R_1 is 3.98 kΩ. Therefore, the obtained value is $R_2 = 1.99$ kΩ. The nearest available resistor values may be chosen for experimental purpose, or series/parallel combinations may be made that are closest to the designed values. For showing actual results of the given example, the circuit is designed using the AD844 sub-circuit in SPICE. The values of resistors used are 4 kΩ and 2 kΩ, while the capacitor value is 100 pF. The signal frequency is taken as 400 kHz. The input and 90° shifted output waveshapes

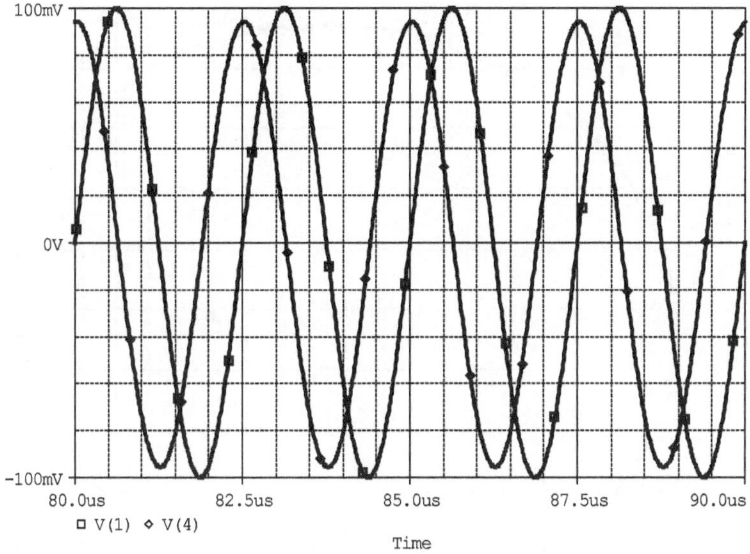

Figure 4.17 Input [V(1)] and output [V(4)] for the design example of circuit in Figure 4.9.

are shown in Figure 4.17. The output is taken from the buffered node in AD844 IC model, as discussed in earlier chapters. The phase difference of 90° between the input and output signal is further shown through the Lissajous pattern in Figure 4.18, where the input signal is taken as x-axis, while the output signal is at y-axis. The circle confirms the 90° phase shift between the input and output signals at the designed pole frequency of 400 kHz. The readers with laboratory access are advised to design the circuit of Figure 4.7. It may be noted that the CCII+-based circuit requires a single AD844 chip, while resistor and capacitor count is similar to the previous example.

4.7 SUPPLEMENTARY READING

The single time constant circuits based on current-mode techniques studied so far are only a small subset of works available in the open literature. The current-mode techniques for such simple electronic functions have been well addressed in hundreds of articles. Besides the simple low-pass and high-pass functions, the more important all-pass function has been the focus of circuit designers. The simplicity of this electronic function coupled with its diverse utility has made it a very useful and popular circuit function. A wealth of articles has thus been contributed on this circuit function. The use of various current conveyor types needs a mention herein for interested readers. The current differencing buffered amplifier (CDBA) is one such active building

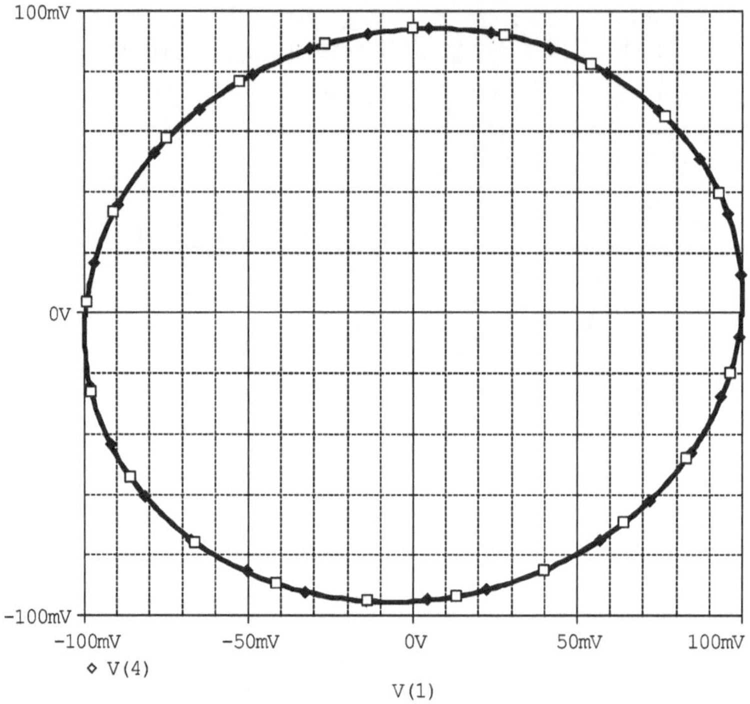

Figure 4.18 Lissajous pattern confirming 90° phase shift between input and output.

block, which has been used effectively for the purpose. The advantage with CDBA is the easy availability of its electronically tuneable version, namely, the current controlled current differencing buffered amplifier (CCCDBA), which allows electronic tuning to the circuits built around it. Another current-mode building block, though not tuneable, has been used for this purpose. The third-generation current conveyor (CCIII) has been found useful for current input, current output circuits. The feature-packed circuits with grounded components and benefitting from high input impedance for voltage-mode operation have been designed using differential voltage current conveyors (DVCCs). The limited tuning ability of DVCC has also been exploited in all-pass filters, where the pole-frequency tuning through bias voltage has been shown. The dual X current conveyor (DXCCII) has been another current-mode building block which has been successfully used for designing all-pass filters with cascading features. The modified version of DXCCII can be found in the literature, where the buffered output stage has been incorporated within a DXCCII. This has made possible the realization of analog circuits with low output impedance, besides high input impedance, making those realizations truly favouring cascade design. A large number of voltage- and current-mode circuits have been designed using a more complex active building block, namely, fully differential current conveyor (FDCCII). The possibility of mixed mode designs, with both voltage and/or current input/output, has been attempted with success using FDCCIIs. There are many interesting circuit designs added to the list, where extra-X current conveyor (EXCCII) and its tuneable version (EXCCCII) have been used for realizing all-pass filters in diverse modes of operation. The benefits of a compact realization with these building blocks have been a feature of significance. These developments have been motivated by the dominance of CMOS technology, with reduced supply voltage operation, wider frequency range of operation, use of fewer components and/or building blocks with fewer transistors, etc. Some of these developments can be easily found for supplementary reading [20–23].

Chapter 5

Higher order analog filters

The coverage on analog circuit design using current-mode techniques has now brought us to a level where more complex analog circuits can be understood. The simple interfacing analog blocks and single time constant circuits have been studied, which form the foundation for higher order circuit design. This chapter deals with the second order filters operating in the voltage and current modes. The mixed mode filter circuits are also covered. The current-mode building blocks used are CCII, CCCII and EXCCII. The electronic tuning aspects are discussed and the filter circuits with such features are also described. The design of higher order filter circuits can then be attempted using second order and first order sections. The realization of circuits using commercially available current-mode ICs is presented, with interesting extension of the study for laboratory exercises. The behaviour of circuits considering the non-ideal active building block is studied. Illustrative examples are given along with real design examples showing graphical results using SPICE.

5.1 SECOND ORDER ANALOG FILTERS

The order of a filter decides the rate at which the gain magnitude and phase response changes with the frequency of the signal. The first order filter circuits are limited by their simplicity to offer slow change in their characteristics, both magnitude and phase. Confining to the magnitude, the first order filters' gain magnitude changes by 20 decibels per decade change in frequency. It is expressed as 20 dB/decade. For the low-pass function it is –20 dB/decade, while for the high-pass function it is +20 dB/decade. The all-pass function is designed to maintain a constant gain magnitude. For the second order analog filters, the changes in characteristics are relatively faster than first order filters. The gain magnitude changes for low-pass and high-pass are –40 dB/decade and +40 dB/decade, respectively. Most of the analog filter theory revolves around the second order filter design, because the higher order filters can be realized using such filters in conjunction with first order sections. This is especially true for active-RC filters designed using

DOI: 10.1201/9781003403111-5

the cascade approach, although various other approaches are available in voluminous subject related texts. The types of second order filters based on the selection or rejection of frequency ranges/bands are of interest. The low-pass, high-pass, band-pass, band-reject and all-pass are five such filter types. As per their names, the frequency selection/rejection is easy to co-relate. The same is illustrated by Figures 5.1, 5.2 and 5.3, which show typical second order gain magnitude functions of five such filters. Figure 5.1 shows low-pass, high-pass and band-pass; Figure 5.2 shows band-reject; and Figure 5.3 shows the all-pass gain magnitude response.

Practice Exercise 5.1: *Based on Figure 5.1, find the amplitude of output signal for three filters, if the input signal is a 1 V peak sine wave of (a) 100 kHz, (b) 1 MHz, (c) 10 MHz.*

Practice Exercise 5.2: *Based on Figure 5.2, repeat problem 5.1 for the band-reject filter.*

Figures 5.1, 5.2 and 5.3 provide the readers a feel of the nature of responses obtained from second order filters. The low-pass allows a range of lower frequency signals, while rejecting higher frequency signals. The high-pass allows higher range of frequency signals, while rejecting lower frequency signals. The band-pass allows a band of frequency signals, while rejecting the out of band frequency signals. The band-reject allows all frequency signals except a band of frequency, which are rejected. The all-pass is a special case allowing all frequency signals with diverse applications in

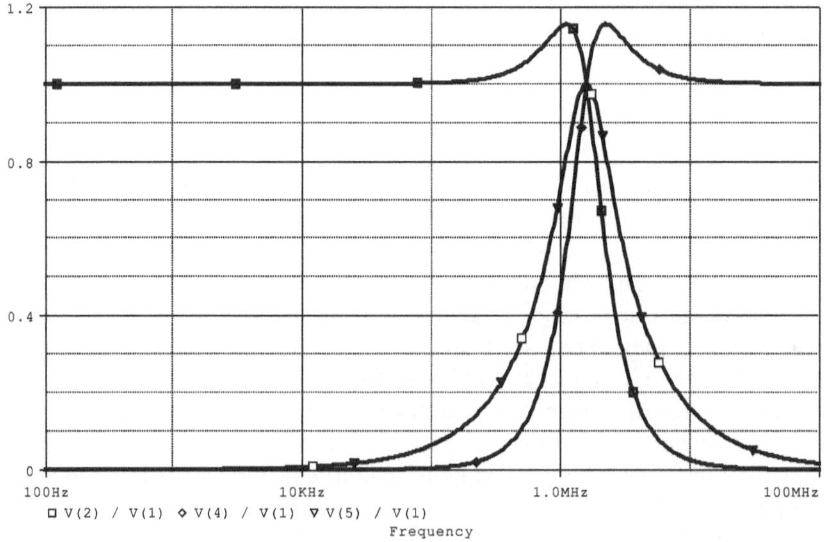

Figure 5.1 Gain magnitude response of low-pass, high-pass and band-pass filter.

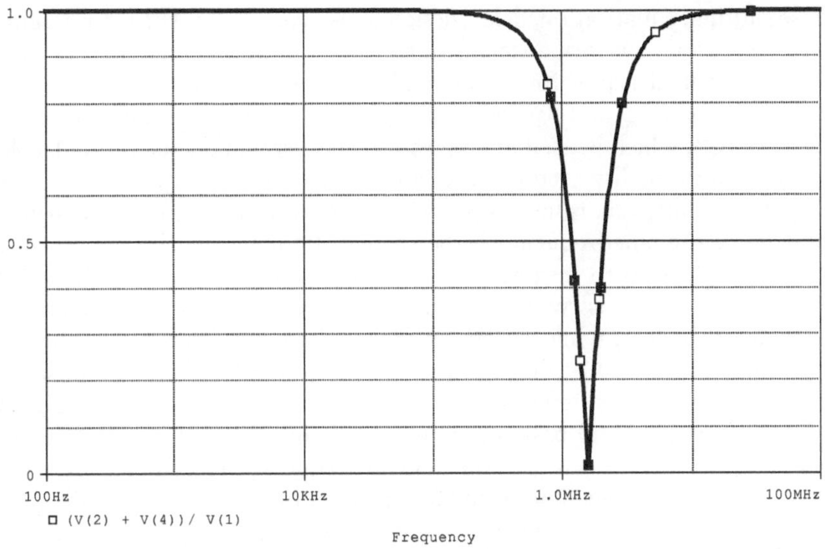

Figure 5.2 Gain magnitude response of band-reject filter.

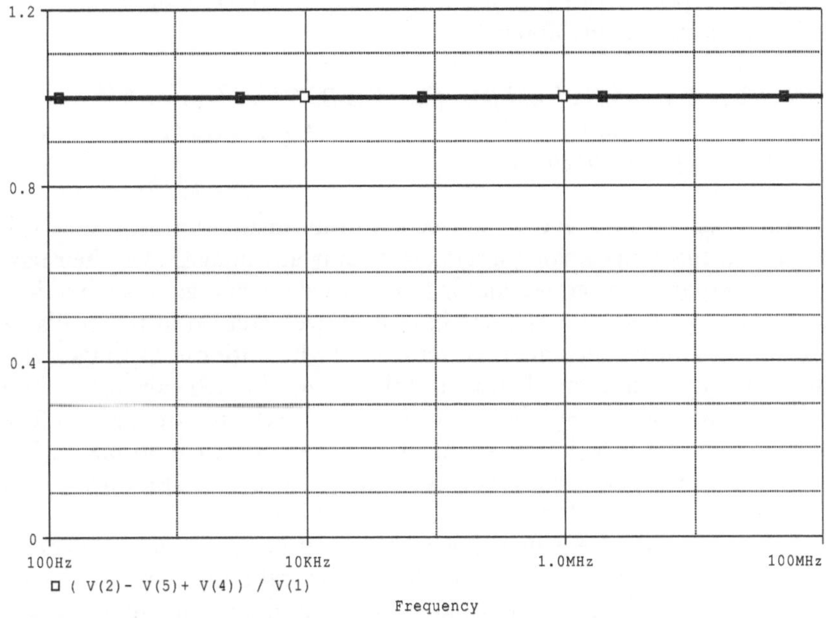

Figure 5.3 Gain magnitude response of all-pass filter.

phase shifting, equalization, delay, design of oscillators, etc. The filter parameters are certain performance measures for a frequency selective circuit. These are defined as pole frequency, quality factor and bandwidth. These are symbolized as ω_o, Q and BW, respectively. The alternative representation for BW is (ω_o/Q), which is itself an expression suggesting that the ratio of pole frequency and quality factor is the filter bandwidth. The quality factor is suggestive of filter sharpness (selectivity) in rejecting/selecting a frequency band, while the pole frequency is the value of frequency where such transition from selection to rejection or vice versa takes place. The pole-Q for low-pass and high-pass filters signifies a bump close to the pole frequency for higher Q values, while selectivity for band-pass and band-reject filters, which increases with Q value. For the band-pass and band-reject filters, the pole frequency is often referred to as the centre frequency, where the gain is either maximum (band-pass) or minimum (band-reject). It may be noted that the pole frequency (ω_o) is the angular frequency (measured in radians per second). The conversion to the frequency (in Hertz) requires a division by 2π, i.e. $f_o = \omega_o/2\pi$. For example, a filter with bandwidth as 1 kHz and quality factor as 10 has the pole frequency of 10 kHz, which is the desired f_o. When expressed in radians per second, the equivalent $\omega_o = 2\pi f_o$. Thus, the angular frequency value obtained is 62.8 krad/s. Another parameter of filter performance measure is the filter passband gain, denoted as H_{type}, where the suffix is one of the filter types, as already mentioned. The abbreviations used for various types are often as *lp* (low-pass), *hp* (high-pass), *bp* (band-pass), *br* (band-reject) and *ap* (all-pass).

Practice Exercise 5.3: *A band-pass filter has BW = 5 kHz and f_o = 10 kHz, find the Q. What would be the new BW, if f_o is changed to 5 kHz, keeping the Q value unchanged.*

With a general introduction to second order analog filters, it is worth noting that the active analog filters offer many advantages over their passive counterparts. The active analog filters based on voltage-mode building blocks like operational amplifiers have been well studied in the available texts on the subject. Such filters suffer from the limitations of voltage operational amplifier, in terms of larger supply voltage, limited gain-bandwidth product, lower slew rate, absence of inherent electronic tuning, etc. The use of current-mode techniques is helpful to overcome such problems. The use of various current conveyor types in place of voltage operational amplifier solves the problem of active analog filter design with lower supply operation, wider frequency range, better accuracy, electronic tunability and integration ease. The sections that follow take the readers to such solutions, with voltage input, voltage output and current input, current output filters using current-mode techniques. Another type of filter to be covered are the mixed mode filters, followed by tuneable filters and other related topics.

5.2 VOLTAGE-MODE CCII-BASED FILTERS

The voltage-mode filters being referred in the header are the filters operating with voltage input signals and providing voltage output signals. These are based on current-mode techniques, thereby employing one or other type of current conveyors. Thus, these filters can also be referred to as voltage-mode filters realized using current-mode techniques. The salient features of such circuits would be known to interested readers. The voltage input, voltage output circuits require high input impedance and low output impedance, so this is one of the desirable features of second order filters operating in the voltage mode. Another feature is the component requirement. A second order filter can be designed using a minimum of two capacitors, if active-RC approach is used. Single active building block-based filters are desirable, but may not necessarily offer other performance feature due to the restriction of external RC network topology to be interconnected to a single active building block. The filters with independent control over filter parameters are of interest because the individual parameter can be controlled through an independent element, which does not affect other filter parameters. The use of passive elements in grounded form is ideal due to several reasons. The use of grounded capacitor is ideal for minimizing the bottom plate parasitic capacitance and favours easy integration of the circuits built using grounded capacitors. The resistors are also preferred in grounded form, because their IC implementation is easier in such form. Another desired feature expected from second order filters is their wide frequency range of operation, as also the linearity aspect. The realization of filter function from a single topology is another criterion of comparing features of distinct circuits. Therefore, single input, multi-output or single output topologies are possible. On the other hand, multi-input, multi-output or single output topologies are also realized, when studying the subject of analog filters. For example a voltage-mode circuit with single input node, providing three filter functions, falls under the single input, three output variety. Similarly, a filter circuit with three input nodes and a single output node falls under the three input, single output variety.

With the preceding background, the first few circuits based on a single current conveyor are shown in Figure 5.4. Each of the three circuits uses a

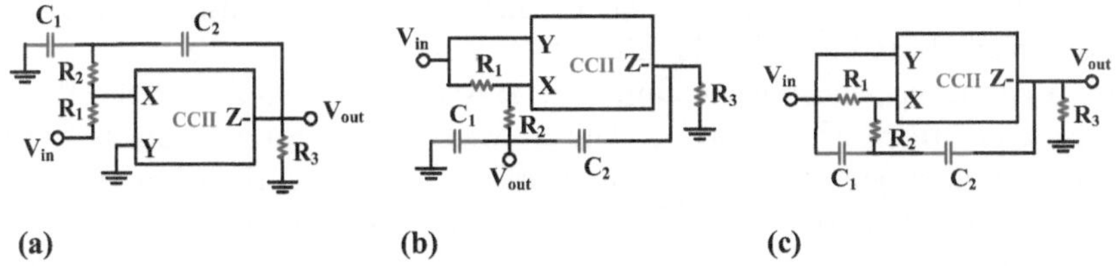

Figure 5.4 Single CCII−-based second order filters.

single CCII– and five passive components. These filter circuits can be derived from the generalized biquadratic filter of ref. [24].

The first circuit is considered for its construction. It is a band-pass filter of second order. The circuit is based on the current follower topology using a CCII–. The voltage input signal appears at one of the resistor's terminal, implying that the input impedance would not be infinite. The output is at the Z– terminal, which is also not a low impedance terminal; hence the output impedance would also not be zero. This aspect is important to note since the circuit is built around a single active building block. The second circuit is of the second order low-pass filter, which also lacks the high input impedance feature. The output signal is again not at low impedance node. The third circuit as shown in Figure 5.4 is a second order high-pass filter. Another common feature of the three circuits is the use of three floating components out of the five passive elements used. However, it is always desired that the filter circuit employs grounded components as far as possible. This aspect is important from the integration perspective of the circuits. Moreover, this aspect is also advantageous in reducing the parasitic effects in the circuits. The three basic filter circuits are thus given in Figure 5.4.

The voltage transfer functions for the three circuits are given in equations 5.1, 5.2 and 5.3, respectively.

$$\frac{v_{\text{out}}}{v_{\text{in}}} = \frac{s/R_1C_1}{s^2 + s\frac{(C_1 + C_2)}{R_3C_1C_2} + \frac{1}{R_2R_3C_1C_2}} \tag{5.1}$$

$$\frac{v_{\text{out}}}{v_{\text{in}}} = \frac{1/R_2R_3C_1C_2}{s^2 + s\frac{(C_1 + C_2)}{R_3C_1C_2} + \frac{1}{R_2R_3C_1C_2}} \tag{5.2}$$

$$\frac{v_{\text{out}}}{v_{\text{in}}} = \frac{s^2}{s^2 + s\frac{(C_1 + C_2)}{R_3C_1C_2} + \frac{1}{R_2R_3C_1C_2}} \tag{5.3}$$

The three transfer functions are now studied further for the filter parameters, namely the pole frequency, quality factor and bandwidth. These terms have already been introduced in the previous section. The three circuits feature a common denominator function. This means the poles of the three transfer functions are the same. The only change in their responses and hence the filter type is due to the numerator function in the three transfer functions. The band-pass transfer function features a zero each at zero and infinite frequency. The low-pass transfer function features a zero at infinity.

The high-pass transfer function exhibits a zero of second order at dc. The filter parameters are now expressed as below.

$$\omega_o = \sqrt{\frac{1}{R_2R_3C_1C_2}}; Q = \frac{1}{C_1 + C_2}\sqrt{\frac{R_3C_1C_2}{R_2}}; \frac{\omega_o}{Q} = \frac{(C_1 + C_2)}{R_3C_1C_2} \tag{5.4}$$

Another parameter of interest is the passband gain, which represents the voltage gain in the passband. The passband gain for the three circuit functions can be expressed as H_{BP}, H_{LP} and H_{HP}, respectively. Their values are given below.

$$H_{BP} = \frac{R_3C_2}{R_1(C_1 + C_2)}; H_{LP} = 1; H_{HP} = 1 \tag{5.5}$$

A detailed discussion on equations 5.4 and 5.5 is now in order. The pole frequency of the three filters can be adjusted through the resistors R_2 and R_3. It is important to note that the capacitor values are normally kept constant and the parameter adjustment is often done through resistive elements. The pole quality factor adjustment requires a variation in the same two resistors and/or capacitors. The filters' bandwidth also depends on the four passive components. This seems to be a problem since the independent adjustment of parameters is difficult. This poses design difficulty in terms of passive elements choice. The passband gains of low-pass and high-pass is unity, but the band-pass filter function can be adjusted for its passband gain independent of any other parameter through R_1. It is important to note that the element R_1 does not appear anywhere else in the filter parameters. The design of the filter circuits discussed above can be further understood with the help of an example. For instance, if the circuit is to be designed using equal-valued capacitors (it is desirable to use equal valued components) for, say, the pole frequency of 100 kHz and pole-Q value of 2, the design begins with a feasible choice of capacitors, say 100 pF (although this value is on the higher side, as far as feasibility is concerned). Seeing the filter parameters' expression, it is now easy to get the values of pole frequency and Q-determining resistors, namely R_2 and R_3. The resulting design equations for equal capacitor design ($C_1 = C_2 = C$) are given below.

$$\omega_o = \frac{1}{C}\sqrt{\frac{1}{R_2R_3}}; Q = \frac{1}{2}\sqrt{\frac{R_3}{R_2}} \tag{5.6}$$

The readers are now encouraged to use the above equation with values of capacitors, pole frequency and quality factor substituted to solve for the unknown resistive elements. It is to be noticed that both resistive ratio and their product need to be set appropriately as per the design specifications, which is a tedious design. The reason is that there is no element in either of the two parameters that can independently set that parameter without affecting the other parameter.

Practice Exercise 5.4: *The circuit of Figure 5.1 is designed using equal capacitors of 50 pF, and equal resistor of 5 kΩ each. Calculate the filter parameters.*

Practice Exercise 5.5: *Design the band-pass filter circuit of Figure 5.2 for Q = 1 and centre frequency of 0.5 MHz.*

The next second order filter circuit to be studied is now shown in Figure 5.5 [25]. The circuit is based on the use of a CCII+ and a CCII–. The circuit employs two capacitors and three resistors. A resistor and capacitor each are grounded, while the rest three passive components not grounded, as shown in Figure 5.5. These would be grounded under certain realizability condition(s), as shall follow in our discussion. As it is evident from the circuit, none of the input signals appears at the *Y* terminal of current conveyor. This means that the circuit does not have high input impedance feature. The output of the circuit is also not at low impedance. However, once the circuit is realized using AD844, the output can be easily tapped from buffered (*W*) output, since the output appears at the *Z* terminal of CCII(2) in the circuit.

Coming to the analysis of the circuit of Figure 5.5, it begins with writing the voltage and current relationships for the CCIIs. The *X*-terminal voltages are to be equated to the *Y*-terminal voltages of the respective CCIIs. The *Z*-terminal currents of the two CCIIs are to be equated to the *X*-terminal currents, with correct polarity, positive for CCII+ and negative for the CCII–. The analysis result is summarized as the equation for output voltage as given below.

$$v_{\text{out}} = \frac{v_1 s^2 + v_2\left(\dfrac{s}{R_1C_1}\right) + v_3\left(1/R_2R_3C_1C_2\right)}{s^2 + s/R_1C_1 + \dfrac{1}{R_2R_3C_1C_2}} \tag{5.7}$$

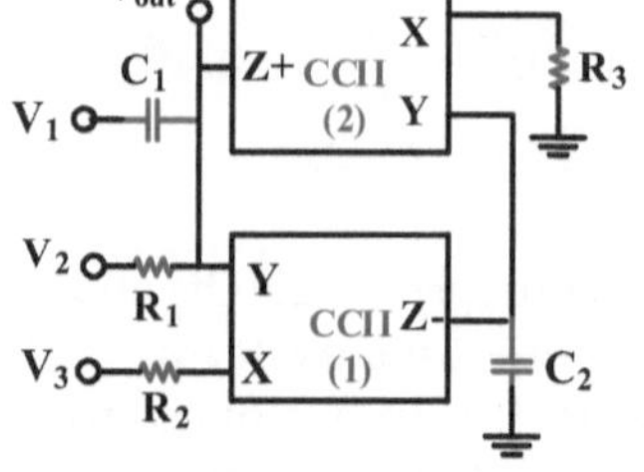

Figure 5.5 Second order filter with three inputs and a single output [25].

Equation (5.7) needs some explanation. The circuit of second order filter being described needs to be specialized in terms of input node selection. Therefore, the output voltage expression would be related to the input, as per the selection of particular input(s). The voltage transfer function is then obtained for the filter functions. Thus, if the input signal is connected at node v_1 and other two input nodes are grounded, then high pass filter function is obtained. The transfer function of the same is given below in equation (5.8).

$$\frac{v_{out}}{v_1} = \frac{s^2}{s^2 + s / R_1C_1 + \dfrac{1}{R_2R_3C_1C_2}} \tag{5.8}$$

The second choice is to connect the input signal to node v_2 and ground the other two input nodes. The transfer function (equation 5.9) thus realized is that of a band-pass filter as given below.

$$\frac{v_{out}}{v_2} = \frac{s / R_1C_1}{s^2 + s / R_1C_1 + \dfrac{1}{R_2R_3C_1C_2}} \tag{5.9}$$

The next choice of input insertion is node v_3, with other two nodes being connected to ground. The filter transfer function now corresponds to a low-pass function (equation 5.10). The same is given below.

$$\frac{v_{out}}{v_3} = \frac{1 / R_2R_3C_1C_2}{s^2 + s / R_1C_1 + \dfrac{1}{R_2R_3C_1C_2}} \tag{5.10}$$

Two more choices exist, namely the insertion of input signal at nodes marked v_1 and v_3, while grounding v_2 and insertion of input signal at all three nodes, with inverted input at v_2. These two choices yield the band-reject and all-pass transfer functions, respectively. Therefore, it is to be noticed that all five standard filter functions of second order can be realized using the circuit. The last two transfer functions are expressed below, where $v_1 = v_3 = v_{in}$ for band-reject function and $v_1 = -v_2 = v_3 = v_{in}$ for the all-pass function, respectively.

$$\frac{v_{out}}{v_{in}} = \frac{s^2 + 1/R_2R_3C_1C_2}{s^2 + s / R_1C_1 + \dfrac{1}{R_2R_3C_1C_2}} \tag{5.11}$$

$$\frac{v_{out}}{v_{in}} = \frac{s^2 - \left(\frac{s}{R_1C_1}\right) + \left(1/_{R_2R_3C_1C_2}\right)}{s^2 + s/R_1C_1 + \frac{1}{R_2R_3C_1C_2}} \tag{5.12}$$

Equations (5.11) and (5.12) correspond to band-reject and all-pass filters, respectively. The filter parameters for the circuit of Figure 5.5 and hence all the five filter functions realized from it are given below.

$$\omega_o = \sqrt{\frac{1}{R_2R_3C_1C_2}};Q = R_1\sqrt{\frac{C_1}{R_2R_3C_2}};\frac{\omega_o}{Q} = \frac{1}{R_1C_1} \tag{5.13}$$

A good feature of the circuit is the possibility of bandwidth variation without affecting the pole frequency. This can be done through resistor R_1, which appears in bandwidth expression, but not in the pole-frequency expression. Therefore, the quality factor can be controlled independent of the pole frequency, and vice versa is also true. The circuit can be designed for high Q values as well, since an independent resistor can vary the quality factor, this feature being especially desirable for band-reject and band-pass functions. This feature is even useful for designing higher order filter using second order filter sections using various filter approximation methods, which are not to be attempted in the present treatment.

Practice Exercise 5.6: *The second order filter of Figure 5.5 is designed using 1 nF capacitors and 5 kΩ resistors. Calculate all filter parameters.*

Practice Exercise 5.7: *Design the second order voltage-mode filter for pole frequency of 1.2 MHz and quality factor of 5. (Hint: Assume capacitors (equal valued, say 50 pF) and use equation (5.13) to find the resistor values)*

5.3 SECOND ORDER CURRENT-MODE FILTERS

The next category of second order filters to be studied under the current-mode circuit design techniques is the ones that operate on input current signals and provide current outputs. It is to be emphasized that current-mode analog building blocks are already being used in the present treatment. Therefore, such filters can be defined as purely current mode in the sense that these are built using current-mode building blocks and process current signals. The desirable feature of such current-mode filters would be similar to any other circuit operating in the current mode. Thus, low input impedance and high output impedance are desired from such filters. The use of grounded passive components is another feature that is important from the IC fabrication perspective and in minimizing the parasitic effects.

For the diversity of analog building block usage, the same to be used in the discussed circuit is a differential voltage current conveyor (DVCC). It is to be reminded to the readers that DVCC has already been introduced in earlier chapters in the book. It is basically a differential version of CCII, with the Y port made dual ended so as to facilitate differential processing of signals. However, in the present context a single ended current input circuit is introduced, with multiple (three) outputs. The circuit is shown in Figure 5.6 [26].

It requires a single DVCC, with two Z stages, two capacitors and two resistors. This component count is a minimum for second order active-RC filter. The input signal is injected through a non-zero impedance (rather high impedance) terminal and the outputs are tapped across passive components, both being the restricting features. However, the simplicity of the circuit, in terms of component requirements, makes it a choice of study at this level. The motive is to introduce the readers to the marvels of such simple designs for current-mode second order filtering functions. The analysis of the circuit is carried out by re-considering the port relationships for a DVCC with two Z stages. These are $v_x = v_{y1} - v_{y2}$; $i_{y1} = i_{y2} = 0$; $i_{z1+} = i_x$; $i_{z2+} = i_x$. The three current-mode transfer functions are given below.

$$\frac{I_{LP}}{I_{in}} = \frac{{}^{1}\!/_{R_1R_2C_1C_2}}{s^2 + s\left(\frac{1}{R_2C_2} - \frac{1}{R_1C_2} + \frac{1}{R_1C_1}\right) + \frac{1}{R_1R_2C_1C_2}} \tag{5.14}$$

$$\frac{I_{BP}}{I_{in}} = \frac{{}^{s}\!/_{R_1C_1}}{s^2 + s\left(\frac{1}{R_2C_2} - \frac{1}{R_1C_2} + \frac{1}{R_1C_1}\right) + \frac{1}{R_1R_2C_1C_2}} \tag{5.15}$$

$$\frac{I_{HP}}{I_{in}} = \frac{s^2}{s^2 + \frac{s}{RC_1} + \frac{1}{R^2C_1C_2}} \quad \left(\text{equal valued resistors}(R)\right) \tag{5.16}$$

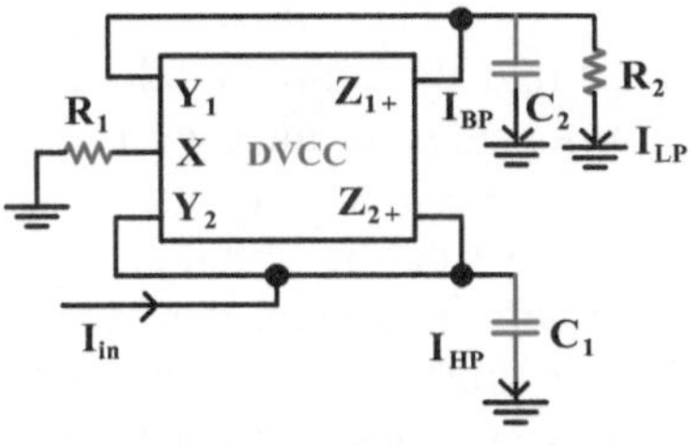

Figure 5.6 DVCC-based current-mode filter with single input and three outputs [26].

The three filter transfer functions given in equations 5.14–5.16 are now described in further detail. The low-pass filter function does not require any condition for realization, and so does the band-pass filter function. However, the high-pass filter function requires a matching condition for its realization. The condition requires the use of equal valued resistors ($R_1 = R_2 = R$). The actual high-pass transfer function, without matching condition is given below, for the readers to appreciate this point. The output has been represented as I_{C1}, rather than I_{HP}, to make the point further clear. The readers are advised to put the condition of equal resistors in the function given below. The resulting high pass transfer function can then be obtained, which is already presented as equation (5.16).

$$\frac{I_{C1}}{I_{in}} = \frac{s^2 + s\left(\frac{1}{R_2C_2} - \frac{1}{R_1C_2}\right)}{s^2 + s\left(\frac{1}{R_2C_2} - \frac{1}{R_1C_2} + \frac{1}{R_1C_1}\right) + \frac{1}{R_1R_2C_1C_2}} \tag{5.17}$$

Therefore, the circuit of Figure 5.6 realizes three basic filter functions operating in the current mode. The filter parameters for the circuit, namely the pole frequency and bandwidth, are given below.

$$\omega_o = \sqrt{\frac{1}{R_1R_2C_1C_2}}; \frac{\omega_o}{Q} = \left[\frac{1}{R_2C_2} - \frac{1}{R_1C_2} + \frac{1}{R_1C_1}\right] \tag{5.18}$$

The quality factor for the circuit can be obtained on dividing the pole-frequency expression by the bandwidth expression. The readers are encouraged to obtain the same. The design of such a filter is relatively difficult, especially for higher Q values. One possible simple design of the circuit is attempted by choosing equal-valued components. This means $R_1 = R_2 = R$ and $C_1 = C_2 = C$. The resulting simplified filter parameters for this design are given below.

$$\omega_o = \frac{1}{RC}; Q = 1; \frac{\omega_o}{Q} = \frac{1}{RC} \tag{5.19}$$

Equation (5.19) suggests that the simplified design results in a filter with fixed Q value of unity. For instance, the filter with equal-valued resistors and capacitors can realize the desired pole frequency and bandwidth, but with the restriction that $Q = 1$. However, the circuit can be designed for Q values other than unity, by modifying the design. For example, if the resistors are assumed to be of equal value (high-pass function even demands this matching), but capacitors are left to the designer's parameter of choice, then several possibilities arise. The expression for the filter parameters then reduces to equation (5.20) as given below.

$$\omega_o = \frac{1}{R}\sqrt{\frac{1}{C_1C_2}}; \frac{\omega_o}{Q} = \frac{1}{RC_1}; Q = \sqrt{\frac{C_1}{C_2}} \tag{5.20}$$

For example, if the filter is now designed for $Q = 5$, it requires the capacitor ratio of 25, which is a high ratio value, keeping in view the integration aspect of circuit components. The choice of say a 10 pF capacitor as C_2 demands the other capacitor (C_1) to be of 250 pF! This is a large-valued capacitor for integration purpose. Theoretically, the filter can be designed for the given pole frequency by appropriate selection of resistor (both of equal value), but practically it is a tough design, which may not be feasible. If the pole frequency of 1 MHz is desired then the value of resistor is found to be 3.18 kΩ. The resistor value is quite feasible, but the capacitor spread (ratio of highest to lowest value) is rather high. Moreover, as already pointed out the value of 250 pF itself is high. This problem is often solved by using capacitance array concept or capacitance multiplier approach. These methods are useful for implementing high capacitor values in integration. The capacitance array concept uses lower value capacitors as an array to realize higher values, while the capacitance multiplier approach employs an active block to realize multiplication factor, which boosts a lower value capacitor. The use of programmable capacitors and multiple layers in fabrication process are also used for realizing large capacitor values in analog circuit design.

Another important feature of the circuit of current-mode second order filter is the output current availability through passive components. This is not a practical way of sensing the output current, unless extra sensing elements are used. Therefore, to solve this problem, the current follower realized using a current conveyor comes to the rescue. At each of the three output nodes, a current follower can be augmented so as to sense the output at high impedance nodes. The readers are encouraged to re-draw the circuit by augmenting current followers, as discussed in earlier chapters.

Practice Exercise 5.8: *The circuit of Figure 5.6 is designed using capacitors of 10 pF and resistors of 2 kΩ. Calculate the pole frequency and filter bandwidth. Re-design the filter for the obtained pole frequency, but quality factor of 2. (Hint: For second part, use equal resistor values, but capacitors are to be selected according to equation (5.20)).*

5.4 ELECTRONICALLY TUNEABLE FILTERS

The second order filters with voltage-mode or current-mode operation as discussed in previous sections are active-RC filters. It is a well-known fact that active-RC filters can be tuned by varying passive components, which is not a desirable feature for integrated filters. The variation in resistors or capacitors for tuning purposes may be a good choice for discrete design, where the graduate learners are required to build circuits on bread-boards. The use of available variable resistors provides an option for the purpose. The availability of varicaps provides another option, when it comes to the tuning of filters through variable capacitors. The active-RC filters can also be made tuneable by using active resistors, wherein the transistor(s)-based resistor equivalent are controlled by a voltage for varying the realized resistance value. The concept of electronically tuneable filters refers to the use of tuneable analog building blocks, which in the present context are current-mode building blocks. This section presents some circuits of second order filters realized using current controlled current conveyors (CCCIIs) and capacitors only. These filters can be called electronically tuneable filters. One such circuit to be studied is shown in Figure 5.7 [27].

The circuit comprises three CCCIIs and two capacitors in grounded form. It is a single input three outputs current-mode filter. The usefulness of CCCIIs for realizing electronically tuneable functions has already been shown in earlier chapters. The circuit has its current input connected at the X terminal, which is low impedance terminal; therefore the circuit enjoys low input impedance feature. The circuit employs one of the CCCIIs as a current follower, with two outputs. Thus, the CCCII with I_{o3} as marked bias current provides two copies of input signal at its $Z+$ outputs. The three shown outputs are available at the Z terminals, which are high impedance terminals; thus, the circuit exhibits the desirable feature of high impedance current outputs. The two capacitors in the circuit are in grounded form, which is again a feature favouring easy integration and ideal for reducing parasitic effects. The circuit analysis can be performed remembering the port relationship of CCCII, already given in earlier chapters. The three transfer functions are expressed as below.

$$\frac{I_{LP}}{I_{\text{in}}} = \frac{1/R_{x1}R_{x2}C_1C_2}{s^2 + s\dfrac{1}{R_{x1}C_2} + \dfrac{1}{R_{x1}R_{x2}C_1C_2}} \tag{5.21}$$

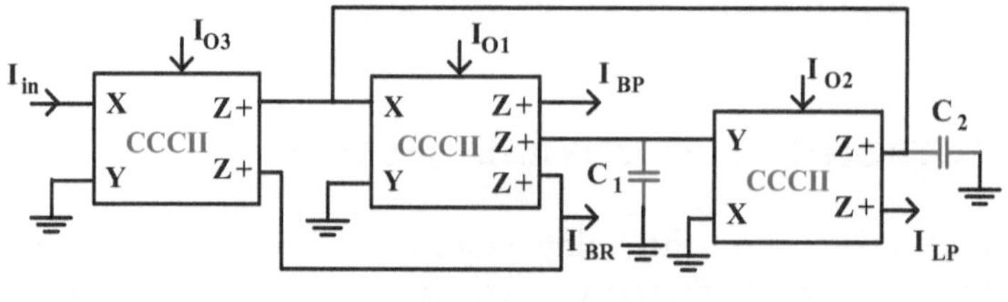

Figure 5.7 Electronically tuneable current-mode filter with single input, three outputs [27].

$$\frac{I_{BP}}{I_{in}} = \frac{s/R_{x1}C_2}{s^2 + s\frac{1}{R_{x1}C_2} + \frac{1}{R_{x1}R_{x2}C_1C_2}} \tag{5.22}$$

$$\frac{I_{BR}}{I_{in}} = \frac{s^2 + \left(1/R_{x1}R_{x2}C_1C_2\right)}{s^2 + s\frac{1}{R_{x1}C_2} + \frac{1}{R_{x1}R_{x2}C_1C_2}} \tag{5.23}$$

The three realized filter transfer functions are of low-pass, band-pass and band-reject filters, as given by equation (5.21)–(5.23), respectively. As already mentioned, only two CCCIIs parameters appear in the transfer functions, the third CCCII being used as a current follower with two outputs for copying the input current. Therefore, the filter parameters depend on R_{x1} and R_{x2}, besides the two capacitors. The expressions for the filter parameters (equation 5.24) are given as below.

$$\omega_o = \sqrt{\frac{1}{R_{x1}R_{x2}C_1C_2}}; Q = \sqrt{\frac{R_{x1}C_2}{R_{x2}C_1}}; \frac{\omega_o}{Q} = \frac{1}{R_{x1}C_2} \tag{5.24}$$

The pole frequency and quality factor can be electronically tuned by controlling the bias current of CCCIIs. For example, if the filter circuit is to be designed for a specific Q value, then the ratio of two bias currents may be maintained, while varying their product to satisfy the given pole frequency. If, the pole-Q value of 2 is desired then the ratio of two intrinsic resistances (R_{x1}/R_{x2}) has to be kept fixed at 4 (for equal capacitor design). Since the bias current for a bipolar CCCII is inversely proportional to its intrinsic resistance ($R_x = V_T/2I_o$, where, V_T is the thermal voltage), this means the ratio of two bias currents (I_{o2}/I_{o1}) has to be chosen as 4. Now for varying the pole frequency, without affecting pole-Q demands keeping the bias current ratio (4, in this example) constant, while varying the individual bias currents. The choice of two bias currents may be chosen as 10 μA and 40 μA. The R_x values for the two bias currents are 1300 Ω and 325 Ω, respectively, assuming 26 mV thermal voltage at room temperature. This choice results in a pole frequency of 2.45 MHz for 100 pF capacitors-based design, with pole-Q as 2. Now keeping the pole-Q same as 2, the pole frequency is varied by changing the bias currents values. If the bias currents are now changed to 20 μA and 80 μA (keeping $Q = 2$), the intrinsic resistance values become 650 Ω and 162 Ω, respectively. The new pole frequency is now found to be 4.9 MHz. This example demonstrates that the filter design is possible by judicious choice of bias currents. The calculations would change if CMOS CCCII is employed for designing the circuit. It may be noted that the intrinsic resistance for a CMOS CCCII is inversely related to the square root of bias current. The readers are advised to obtain the filter parameters in terms of the bias currents of CCCII, employing CMOS technology. This can be done using the intrinsic resistance expression for a CMOS CCCII.

Practice Exercise 5.9: *Design the current-mode filter of Figure 5.7 for varying pole frequencies and fixed quality factor (Q = 1), using 50 pF capacitors. The pole frequency is to be varied from 100 kHz to 1 MHz.*

Practice Exercise 5.10: *The current-mode filter of Figure 5.7 was designed for equal bias currents of CCCIIs. The designed pole frequency desired was 2 MHz. Find the required value of capacitors, if a Q value of 5 is required. (Hint: Use ratio of capacitors to satisfy the Q value)*

The three-output circuit of Figure 5.7 can be extended for the remaining filter functions, namely, high-pass and all-pass. The availability of current outputs at the *Z* terminals provides designers with the ease of extending the circuit for more responses. The high impedance current outputs can be connected together for summing operation, a fact which is well known to the readers of this text from earlier chapters. The flexibility of CCCII for augmenting additional *Z* stages is another design ease with the approach. The *Z* stages with desired polarity (positive or negative) can be added to the CCCII circuitry. The additional positive stage provides a copy of *X*-terminal current with positive polarity, while a negative stage provides an inverted copy of *X*-terminal current. This feature can be exploited to extend the circuit, with additional *Z*+ and/or *Z*– stages, so as to provide additional filter functions. For example, the addition of a *Z*– stage in CCCII (with bias current I_{o2}) yields a response, which is inverted version of current available at its *Z*+ stage. Thus, it would provide a $-I_{LP}$ response. Now, if this output is shorted with the I_{BR} output node, a current summation ($I_{BR} - I_{LP}$) occurs, resulting in an output which is high-pass in nature! The following transfer function (equation 5.25) is obtained as a result of this modification.

$$\frac{I_{HP}}{I_{in}} = \frac{s^2}{s^2 + s\dfrac{1}{R_{x1}C_2} + \dfrac{1}{R_{x1}R_{x2}C_1C_2}} \tag{5.25}$$

Another circuit modification can be made to obtain the all-pass response. The CCCII with bias current I_{o1} can be modified with an additional *Z*– stage. The band-pass output is available from *Z*+ stage of this CCCII. The additional *Z*– stage results in the realization of a response, which is $-I_{BP}$. Now shorting this output with I_{BR} yields the all-pass response. The resulting transfer function is as equation (5.26).

$$\frac{I_{AP}}{I_{in}} = \frac{s^2 - s\dfrac{1}{R_{x1}C_2} + \left(1/R_{x1}R_{x2}C_1C_2\right)}{s^2 + s\dfrac{1}{R_{x1}C_2} + \dfrac{1}{R_{x1}R_{x2}C_1C_2}} \tag{5.26}$$

With such modifications, the circuit's utility can be extended. The readers are now encouraged to convert the circuit of Figure 5.7 into a single input, five outputs filter. It may be noted that additional Z stages are to be included in CCCII implementation. It would be an interesting project problem at the Master's level. The complete layout and fabrication aspect could become a research problem at the doctoral level. For the undergraduate learners, the simulation and analysis of the circuit can best be carried out using available CAD tools. The readers with circuit design as a hobby can assemble the circuit using off-the-shelf components! However, such designs are best suited for monolithic integration. The reasons why this filter suits integration are the features offered by the circuit. The use of grounded capacitors, current-mode operation with low input and high output impedance and electronic tuning of filter parameters makes the circuit apt for integration. The cascading feature of the circuit without using additional current followers further allows for design of higher order filters. Before proceeding to other higher order filters, the section that follows next covers high input impedance voltage-mode filters.

5.5 VOLTAGE-MODE FILTERS WITH HIGH INPUT IMPEDANCE

The problem with the filters operating in voltage mode studied thus far is the absence of un-conditional high input impedance. This feature is of special interest for practical reasons, when the second order filters are used within a system and their connections with other blocks needs appropriate input and output impedances. The current-mode techniques for designing such filters are extended further with another circuit, which fulfils the high input impedance requirement, without any conditions. The circuit is designed using DVCCs and passive components, falling under active-RC approach. The circuit is shown in Figure 5.8 and its description is as follows [28]. It is based on three DVCCs and five passive elements. The circuit is voltage input driven and provides five simultaneous outputs, hence may be called a single input, five outputs filter circuit. All except one passive component are grounded in the circuit and the input impedance is infinite. It is worth pointing out that filter circuits providing all standard filter functions are called universal filters.

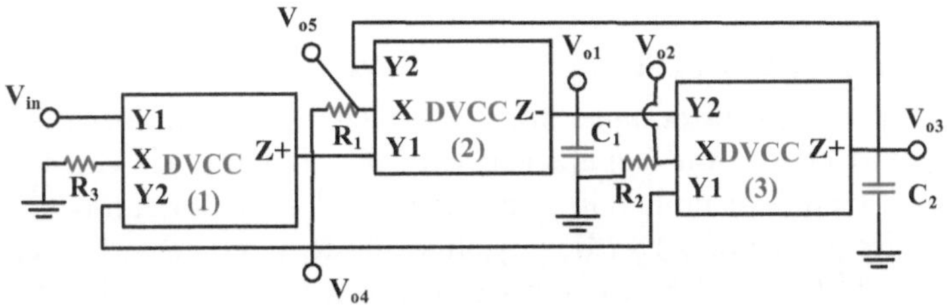

Figure 5.8 Single input five outputs filter using DVCCs [28].

It simply means that the same topology is capable of providing all standard filter transfer functions. The circuit transfer functions are given as below.

$$\frac{v_{o1}}{v_{in}} = \frac{1/R_1R_2C_1C_2}{s^2 + s\dfrac{R_3}{R_1R_2C_2} + \dfrac{1}{R_1R_2C_1C_2}} \tag{5.27}$$

$$\frac{v_{o2}}{v_{in}} = \frac{s^2}{s^2 + s\dfrac{R_3}{R_1R_2C_2} + \dfrac{1}{R_1R_2C_1C_2}} \tag{5.28}$$

$$\frac{v_{o3}}{v_{in}} = \frac{\dfrac{s}{R_2C_2}}{s^2 + s\dfrac{R_3}{R_1R_2C_2} + \dfrac{1}{R_1R_2C_1C_2}} \tag{5.29}$$

$$\frac{v_{o4}}{v_{in}} = \frac{s^2 + 1/R_1R_2C_1C_2}{s^2 + s\dfrac{R_3}{R_1R_2C_2} + \dfrac{1}{R_1R_2C_1C_2}} \tag{5.30}$$

$$\frac{v_{o5}}{v_{in}} = \frac{s^2 - \dfrac{s}{R_2C_2} + 1/R_1R_2C_1C_2}{s^2 + s\dfrac{R_3}{R_1R_2C_2} + \dfrac{1}{R_1R_2C_1C_2}} \tag{5.31}$$

Equations 5.27 to 5.31 are the transfer functions for low-pass, high-pass, band-pass, band-reject and all-pass filters, respectively. The circuit is unique in the sense that all five standard filtering functions are simultaneously realized from it without any matching conditions. The filter parameters are given below.

$$\omega_o = \sqrt{\frac{1}{R_1R_2C_1C_2}};\ \frac{\omega_o}{Q} = \frac{R_3}{R_1R_2C_2};\ Q = \frac{1}{R_3}\sqrt{\frac{R_1R_2C_2}{C_1}} \tag{5.32}$$

The pole-Q can be varied by R_3 without affecting the pole frequency. The filter circuit can thus be designed for high Q values as well, since there is no interaction of pole-Q varying element with the pole frequency. One simple design for $Q = 1$ is given herein as an exercise. The circuit can be designed for a pole frequency of 3.18 MHz by choosing $R_1 = R_2 = R_3 = 5$ kΩ and $C_1 = C_2 = 10$ pF. The readers are advised to check the design using equation (5.32). However, of more importance is the variation of Q or high Q design, which is also possible with this circuit. For the given values of R_1, R_2, C_1 and C_2, the pole-Q variation is possible by controlling R_3. For example, if $Q = 5$ is desired then the value of R_3 need to be appropriately chosen. A simple analysis yields the value of R_3 as 1 kΩ. Now the Q value can be further increased to 10 by reducing R_3 to 0.5 kΩ. This design flexibility was missing in some of the circuits studied earlier in this chapter.

Practice Exercise 5.11: *Design the universal filter of Figure 5.8 for a pole frequency of 5 MHz and pole-Q value as 2. Simulate the filter using DVCC model given in earlier chapters and verify your design.*

Continuing the critical study on the universal filter, it is important to note that the accuracy of filter parameters depends on the circuit topology and parasitic effects. It also depends on the active building block characteristics. These concerns can be discussed one by one. The circuit of the universal filter based on DVCC uses resistive terminations at the X terminals of DVCCs. There are capacitors at Y-Z shorted nodes. Both these topological features are favourable from the parasitic viewpoint. As it is well known that a current conveyor X terminal exhibits finite X-terminal resistance, so any resistive termination at X favours merger of intrinsic resistance with external resistor. The designer must choose external resistor value such that it is large in comparison with the DVCC intrinsic X-terminal resistance. The Y-Z shorted terminals with external capacitors are also considered favouring parasitic absorption in the circuit. The Y and Z terminal of a DVCC are characterized by parasitic capacitances, which appear in shunt with externally terminated grounded capacitors in the circuit. The designer must choose external capacitors which are large as compared to expected DVCC parasitic capacitances at the Y and Z terminals. This way errors in the realized filter parameters can be minimized. The other concern is of DVCC voltage and current transfer gains, which depends on the technology choice, supply voltage and CMOS implementation of DVCC. An accurate DVCC implementation is helpful in reducing errors due to any in-accuracy in voltage and current transfer gains. The frequency range of operation also depends on the CMOS circuitry and technology node used to implement the DVCC. The higher frequency operation depends on the DVCC's usability at higher frequencies. It is directly related to the lower –3 dB values of voltage and current transfer gains.

5.6 HIGHER ORDER FILTER DESIGN

The design of analog filters using current-mode techniques is now extended to higher order filters. The first and second order filters are limited to the stopband attenuation rates which are –20 and –40 dB per decade, respectively. The sharp cut-off frequency is often the need of analog filters in many communication applications. The ideal brick wall characteristic for low-pass or high-pass filters is hard to achieve using electronic components and building blocks. The feasible nature of transition at cut-off frequency is a brick wall approximation only. The need of highly selective band-reject and band-pass filters demands the use of higher order designs. The all-pass filters of higher order are needed for phase equalizations and delay applications. Therefore, the design of such filters is carried out using various standard approaches. The ladder filters are often used as prototypes, wherein the inductors are replaced by active-RC equivalents to realize higher order transfer functions. This approach demands the use of current-mode building blocks to replace inductors by their current conveyor-based simulators. Two types of networks can be obtained using this approach. The first one is based on replacing an inductor by current conveyor and RC network. The resulting current-mode filter would be called an active-RC higher order filter using the current-mode technique. The second approach is to use tuneable current conveyors. This requires the use of current controlled conveyors and capacitors only. The resulting network is then called an active-C current-mode higher order filter, with the advantage of electronic tuning of filter parameters. The cascade design approach relies on first and second order filter sections to realize higher order filters. This approach is relatively simple but requires appropriate input and output impedances for easy cascading of various lower order sections to realize higher order transfer functions. There can be two types of higher order filters under this category. The voltage transfer functions and current transfer functions can be realized using lower order sections, thereby giving rise to higher order voltage-mode filters and current-mode filters using current-mode techniques. The design of higher order filters using current-mode techniques based on various approximation methods is also possible using standard second order and first order sections. Therefore, it seems that the development of current-mode techniques for higher order filter design necessitates the availability of inductance simulators and lower order filter sections. The topic would not be elaborated further herein, but the interested readers can explore the area from published literature [29–31].

5.7 MIXED-MODE FILTER CIRCUITS

The operation of analog filters discussed so far has been restricted to realizing either the voltage transfer function or the current transfer function. The filters so realized have thus been referred to as voltage- and current-mode filters using current-mode techniques. Another type of analog filter can be designed using current-mode techniques. These are termed as mixed mode filters. The system-level design often requires change of signal variables from voltage to current and vice versa. These requirements are fulfilled by employing separate voltage to current converter and current to voltage converter circuits. The treatment of such circuits has been given earlier in Chapter 2. However, a more realistic approach for fulfilling this requirement could be designing circuits with voltage/current input and voltage/current outputs. This means that the same circuit can operate on voltage and/or current input signal and provide current and/or voltage output signal. Such filters are called mixed mode filters. These filters provide the flexibility of mode of operation, while also eliminating voltage to current and current to voltage converter blocks. One such circuit is described herein, which is based on CCCIIs and capacitors, hence falling under the active-C category. The electronically tuneable circuit is shown in Figure 5.9 [32]. The circuit requires four CCCIIs and two capacitors to realize second order transfer functions. Both voltage and current input signals can be applied to the circuit and both voltage and current outputs are obtainable.

The circuit operation can be explained by assuming one type of input at a time. This means that either a current input or a voltage input is applied at a time. Each of the two cases results in two types of outputs. These are current and voltage outputs, both being obtained simultaneously. So let us first consider the current input topology, grounding the voltage input node ($V_{\text{in}} = 0$). For this case, the following transfer functions are realized.

$$\frac{I_{HP}}{I_{\text{in}}} = \frac{-\left(\frac{R_{x4}}{R_{x1}}\right)s^2}{s^2 + s\frac{2}{R_{x1}C_1} + \frac{2R_{x4}}{R_{x1}R_{x2}R_{x3}C_1C_2}} \tag{5.33}$$

$$\frac{I_{BP}}{I_{\text{in}}} = \frac{\left(\frac{2R_{x4}}{R_{x1}R_{x2}C_1}\right)s}{s^2 + s\frac{2}{R_{x1}C_1} + \frac{2R_{x4}}{R_{x1}R_{x2}R_{x3}C_1C_2}} \tag{5.34}$$

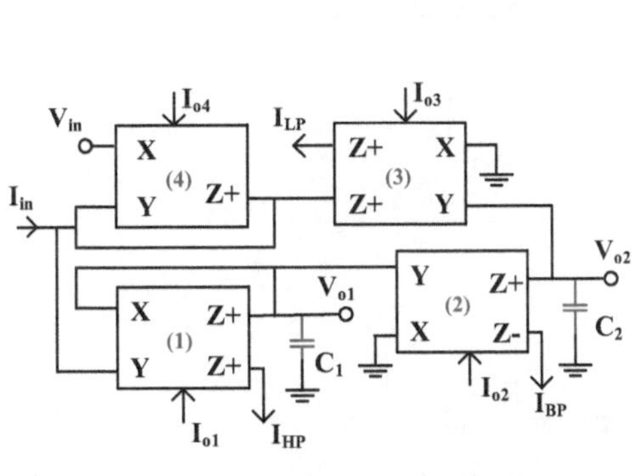

Figure 5.9 Electronically tuneable mixed mode filter circuit [32].

$$\frac{I_{LP}}{I_{\text{in}}}=\frac{-\left(\dfrac{2R_{x4}}{R_{x1}R_{x2}R_{x3}C_1C_2}\right)}{s^2+s\dfrac{2}{R_{x1}C_1}+\dfrac{2R_{x4}}{R_{x1}R_{x2}R_{x3}C_1C_2}} \tag{5.35}$$

$$\frac{V_{o1}}{I_{\text{in}}}=\frac{-\left(\dfrac{R_{x4}}{R_{x1}C_1}\right)s}{s^2+s\dfrac{2}{R_{x1}C_1}+\dfrac{2R_{x4}}{R_{x1}R_{x2}R_{x3}C_1C_2}} \tag{5.36}$$

$$\frac{V_{o2}}{I_{\text{in}}}=\frac{-\left(\dfrac{2R_{x4}}{R_{x1}R_{x2}C_1C_2}\right)}{s^2+s\dfrac{2}{R_{x1}C_1}+\dfrac{2R_{x4}}{R_{x1}R_{x2}R_{x3}C_1C_2}} \tag{5.37}$$

The above equations are now interpreted one by one. The equations (5.33–5.35) are trivial to understand in the sense that the current transfer functions realized are similar to many other filter functions already covered until now. The three filter functions thus realized are high-pass, band-pass and low-pass, respectively. The last two equations are for the voltage outputs, with current as the input signal. Therefore, equations (5.36–5.37) represent the band-pass and low-pass functions. The current input condition realizes five functions in the form of three current outputs and two voltage outputs. The next case would be when the voltage input signal is applied to the circuit and current input is grounded. The resulting filter functions are expressed as below.

$$\frac{I_{HP}}{V_{\text{in}}}=\frac{-\left(\dfrac{1}{R_{x1}}\right)s^2}{s^2+s\dfrac{2}{R_{x1}C_1}+\dfrac{2R_{x4}}{R_{x1}R_{x2}R_{x3}C_1C_2}} \tag{5.38}$$

$$\frac{I_{BP}}{V_{\text{in}}}=\frac{\left(\dfrac{2}{R_{x1}R_{x2}C_1}\right)s}{s^2+s\dfrac{2}{R_{x1}C_1}+\dfrac{2R_{x4}}{R_{x1}R_{x2}R_{x3}C_1C_2}} \tag{5.39}$$

$$\frac{I_{LP}}{V_{\text{in}}}=\frac{-\left(\dfrac{2}{R_{x1}R_{x2}R_{x3}C_1C_2}\right)}{s^2+s\dfrac{2}{R_{x1}C_1}+\dfrac{2R_{x4}}{R_{x1}R_{x2}R_{x3}C_1C_2}} \tag{5.40}$$

$$\frac{V_{o1}}{V_{in}} = \frac{-\left(\frac{2}{R_{x1}C_1}\right)s}{s^2 + s\frac{2}{R_{x1}C_1} + \frac{2R_{x4}}{R_{x1}R_{x2}R_{x3}C_1C_2}} \tag{5.41}$$

$$\frac{V_{o2}}{V_{in}} = \frac{-\left(\frac{2}{R_{x1}R_{x2}C_1C_2}\right)}{s^2 + s\frac{2}{R_{x1}C_1} + \frac{2R_{x4}}{R_{x1}R_{x2}R_{x3}C_1C_2}} \tag{5.42}$$

The above equations (5.38 to 5.42) represent the functions realized with voltage input to the circuit. The first three transfer functions are for three standard current output filters, while the last two functions are for band-pass and low-pass filters. The filter parameters for the mixed mode filter are now given below.

$$\omega_o = \sqrt{\frac{2R_{x4}}{R_{x1}R_{x2}R_{x3}C_1C_2}}; Q = \sqrt{\frac{R_{x1}R_{x4}C_1}{2R_{x2}R_{x3}C_2}}; \frac{\omega_o}{Q} = \frac{2}{R_{x1}C_1} \tag{5.43}$$

Equation (5.43) lists the three filter parameters, namely, pole frequency, quality factor and bandwidth, with certain options of independent tuning of one parameter without affecting the other. The readers are encouraged to find these options of tuning the filter circuit as a simple exercise. Of more importance is the aspect of mixed mode operation, which makes this circuit different from the rest of the circuits discussed so far in this text. Another aspect worth exploring is the filter gain expressions for each realized function. The readers can attempt it as another exercise to complete the circuit analysis. It is worth pointing out that all the current outputs in the circuit are at high impedance nodes, hence fulfilling the cascading feature, without additional buffers. However, the voltage outputs are not at desirable low impedance nodes, thus necessitating additional voltage followers for cascading.

The design of mixed mode filter as reported in ref. [32] is given here as an example. The selection of capacitors as 10 nF (although these values are really large), and bias currents as $I_{o1} = I_{o2} = I_{o4} = 50$ μA and $I_{o3} = 100$ μA yields the pole frequency as 115 kHz and $Q = 1$, for bipolar implementation of CCCII. The tuning of pole-Q may be carried out using capacitor ratio, say C_1/C_2 as 5, and varying I_{o3}. The other three bias currents may be set to equal values. The circuit may be designed using CMOS implementation of CCCII and its performance evaluated as an exercise by interested readers. The capacitors' choice may be in tens of picofarads for making the circuit integrable. Another aspect related to the mixed mode filter is its extension for realizing the remaining filter functions. The current outputs are available for high-pass, band-pass and low-pass functions. It is worth noting that low-pass function and high-pass function are available in inverting mode, while the band-pass function is non-inverting (equations 5.34–5.35). Therefore, connecting the low-pass and high-pass current outputs yields an inverting band-reject function. Similarly, connecting the low-pass, band-pass and high-pass current outputs yields an inverting all-pass filter function. The additional Z stages of correct polarity may be implemented for simultaneous availability of all five filter functions from the circuit. The voltage input, current output functions for band-reject and all-pass filters are similarly obtained. This can be easily verified from equations (5.38–5.40).

Practice Exercise 5.12: *Design a mixed mode filter for a pole Q of 1 and pole frequency of 1 MHz. Compare the designs for bipolar and CMOS CCCIIs. (Hint: Use R_x expressions for bipolar and CMOS CCCIIs from earlier chapters)*

5.8 SHADOW FILTERS

A relatively recent type of electronically tuneable filters is referred to as shadow filters. This is because the shadow filter possesses the ability to electronically modify the filter's parameters, for example, the centre frequency or quality factor externally without involving the filter core circuitry. This is achieved by an externally connected amplifier, without affecting the active and passive components of the filter. The gain of an external amplifier, connected in a feedback path is varied so as to control a filter parameter of interest [33, 34]. Depending upon the value of gain of the feedback amplifier, the filter characteristics can be varied. The realization and design of active filters is dictated by the specific application requirements, which may be categorized in several ways. Besides the passband and stopband attenuation criteria, the filter characteristics, the approximation type and the order of filter often become the guiding factors for appropriate choice, out of the various types discussed in preceding sections. Nonetheless, the tuning requirements are mandatory for desired applications in each of the above realizations. Applications based on the usable frequencies may range from bio-medical, audio, video and radio frequencies. Accurate setting of filter parameters, as per the specifications, is a challenging job of circuit designer. The literature on filter theory and realization methods, along with tuning requirements, has been published, without much attention of the practicality aspects for certain emerging and future applications. One such area is the applications based on Internet-of-Things (IOT). The utility of filters in IOT is expected to be felt in sensing/measuring/pre-processing to communication and postprocessing/interpreting the information. The challenges in designing filters for this area is going to be quite demanding as far as designer's skills and design methods are concerned. The topic of tuning of filters is also expected to be equally challenging for emerging areas like IOT. Whereas tuning using traditional methods of varying a component or the control current/voltage will require several trade-offs and feasibility issues, more recent methods like shadow filters will be under close scrutiny for such real applications. The problems to be addressed for employing shadow filter concept are too many, which make them a cautious choice. The first issue in using shadow filters is the increased circuit complexity. This one issue leads to the problems of chip area and power consumption, which are two leading optimization requirements for IOT success. Another requirement for IOT applications is the reliability of the circuits. The shadow filter may come handy to fulfil this aspect successfully. The traditional tuning methods of varying a component of filter core results in continuous variations in currents/voltages within the devices used for implementing the circuit. These changes are not desirable both for long-term usage and reliable operations. The filter circuit becomes more prone to failures induced by such changes in operating conditions, as a result of current/voltage variations. The use of shadow filter may isolate the filter core from such undesired changes, imparting better reliability to the circuits. However, the use of external amplifier for control/tuning of parameters would still require us to address the reliability issues. This seems to be a blessing in disguise, by providing isolation to the actual filter core. The area and power issues need to be taken care of if shadow filter concept is to find real applications in the IOT-based future world.

5.9 SUMMARIZED CONCLUSION

The chapter focussed on the design of second order filters using current-mode techniques. The introductory material on the topic was covered. The filters operating in the voltage mode, current mode and mixed mode were studied. The CCII, DVCC and CCCIIs were mainly employed for presenting the contents. The single input and single output voltage-mode filters using CCII were studied. The filters built using a single active building block provide compact circuit realization. The multi-input and single output filters of both voltage-mode and current-mode variety were covered. The voltage-mode filters with appropriate input impedance were of special interest due to their easy cascading feature. The mixed mode filters were studied with the benefits of their versatile operation modes. The electronically tuneable filters provided the advantageous features of integration and electronic control over filter parameters. CCII-based circuits can be experimentally realized using AD-844 ICs. DVCC-based circuits can be tested using three ICs per DVCC. The higher order filters studied offer solutions for strict passband and stopband requirements. The salient features of the current-mode techniques ensure filter circuit realizations with simple circuitry, extended bandwidth, low supply operation and good dynamic range. The readers are encouraged to actually design higher order filters of order greater than two using the chapter's contents.

Chapter 6

Non-linear applications

The use of current-mode techniques for realizing linear analog signal processing functions has been so far covered in the preceding chapters. The current-mode building blocks have the inherent advantages of wider frequency range of operation and higher slew rates. These features make the current-mode building blocks a good choice for a variety of non-linear analog signal processing functions. This chapter introduces several such electronic functions, which were traditionally designed using voltage operational amplifiers. The functions presented herein include simple comparators and precision rectifiers. The use of current conveyors in designing detectors is given. The more complex circuits of modulators are discussed. The current-mode logic gates realization is also presented. The current-mode building blocks used are CCII, DVCC, EXCCII and other current conveyor variants. The compatibility of such circuits for IC realization is mentioned for laboratory exercises, while CMOS compatibility makes the designs suitable for future integration.

6.1 COMPARATORS

The comparison of two signals is one of the standard functions encountered in electronic circuits and system design applications. The operational amplifier-based designs are limited by the finite gain-bandwidth product and limited slew rates of operational amplifiers. Thus, the circuits realized are not suited for high-speed applications. The use of current conveyors can overcome these limitations. The CCII itself is an active building block, which inherently allows for easy comparison of two signals. Let a CCII be driven by two signals v_1 and v_2 connected to its Y and X terminals, respectively. The intrinsic resistance at the X terminal was earlier modelled as R_x. The current developed at the X terminal therefore becomes $(v_1 - v_2)/R_x$. It would be well known to the readers who have gone through the earlier chapters that the output current at CCII's Z terminal is equal to the X-terminal current. Therefore, the output current is same as the X-terminal current, which is $(v_1 - v_2)/R_x$. A closer look into the current expression suggests that the

DOI: 10.1201/9781003403111-6

difference of inputs ($v_1 - v_2$) decides the polarity of output current (i_z). For $v_1 > v_2$, i_z is positive, while for $v_1 < v_2$, i_z is negative. This becomes the basis of comparison of the two signals. Now the Z terminal of CCII is characterized by a high output resistance (R_z). When the current i_z flows through this resistance (inherent), the voltage developed at the $Z-$ terminal is of positive or negative value, depending on the two signals v_1 and v_2. Since the output resistance R_z is quite high (it is the output resistance of two transistors comprising the stage), while R_x is quite small, the output voltage (v_z) appears with a gain of R_z/R_x, which is quite large. Since small signals are being compared, and the low supply voltages are being used for designing the CCII, the output voltage simply saturates to a positive or negative saturation voltage. This may be denoted as V_{sat} and $-V_{sat}$. The positive saturation voltage corresponds to $v_1 > v_2$, while the negative saturation voltage corresponds to $v_1 < v_2$. Therefore, the CCII works as a comparator generating a fix output in response to the signals being compared. An example of a comparator designed using a bipolar CCII, biased at ±2.5 V, with two signals v_1 and v_2 applied at the Y and X terminals, respectively, is illustrated by actual SPICE results as shown in Figure 6.1. The output voltage at the Z terminal is found to attain positive or negative saturation values for the inputs being compared. The example shown here compares two ac signals. In the next

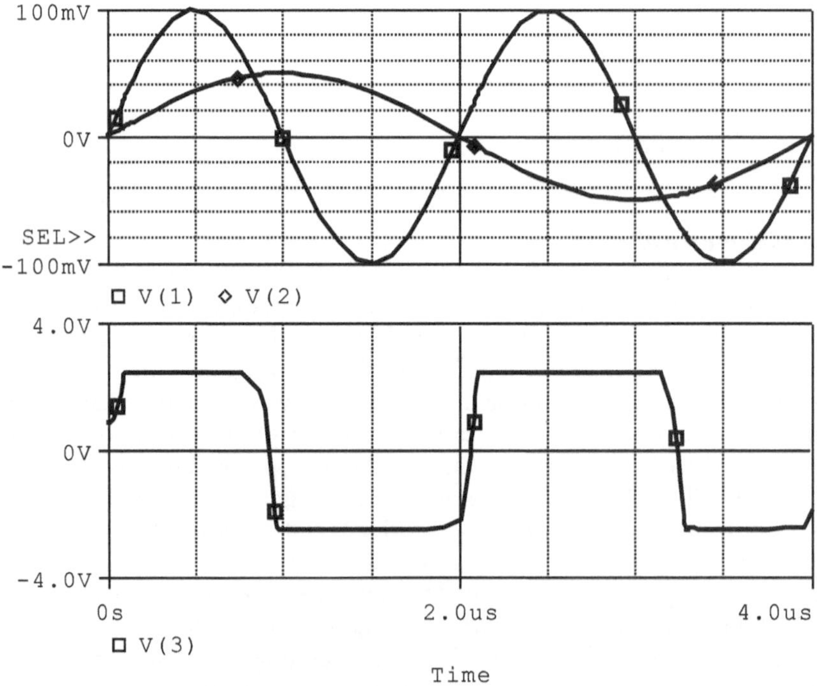

Figure 6.1 Two ac signals and output of a CCII-based comparator.

example the second signal (v_2) is assumed to be a dc voltage, as shown in Figure 6.2, where the output is positive for $v_1 > v_2$, and negative otherwise. The sensitivity of the CCII for comparing small nature of signals can be further emphasized from the dc transfer curve, where the output is plotted with respect to the difference input, which is being taken as dc signal. The curve of Figure 6.3 shows the sharp transition from positive to negative value of the output near zero difference input. The positive saturation occurs for $v_1 > v_2$, while negative saturation occurs for $v_1 < v_2$. One special case of the comparator designed using a CCII is when the input signal zero crossings are to be detected. The reference signal (v_2) is connected to ground for this purpose. This connection is used for detecting zero crossings of the signal connected at v_1. For instance, if v_1 is a sinusoidal signal, then the output voltage at the Z terminal of the CCII is a square wave. Besides zero-crossing detection, the circuit arrangement also works as a simple sinusoidal to square converter. The result using a CCII is shown in Figure 6.4.

In a similar manner as shown above, the other current-mode building blocks can be used as simple comparators. The next example is of a DVCC, which has inbuilt differential processing capability. This feature can be used to design comparators. The DVCC has differential Y input but single ended X terminal. Therefore, the two signals being compared can be connected to

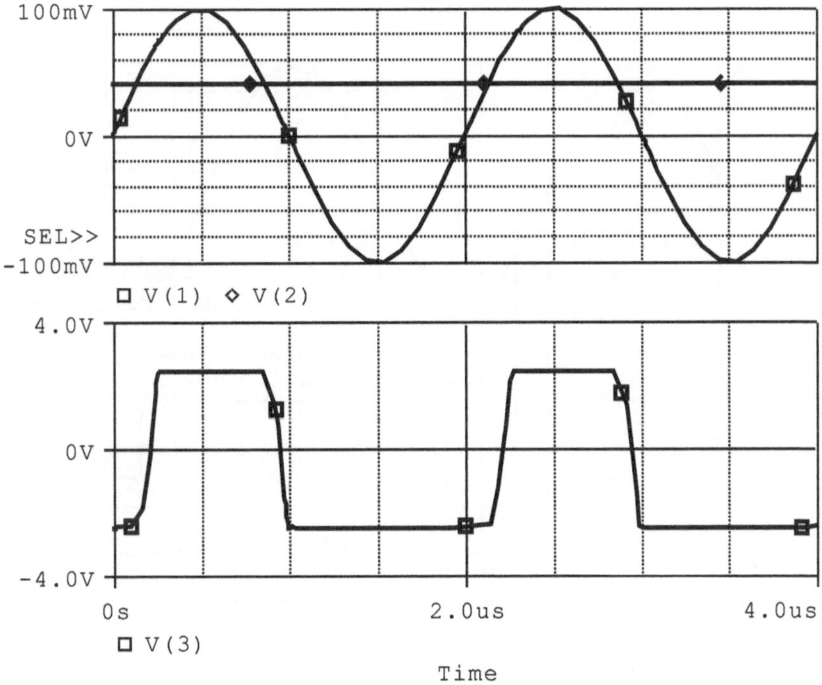

Figure 6.2 The output [v(3)] for v(1) as ac signal and v(2) as dc signal.

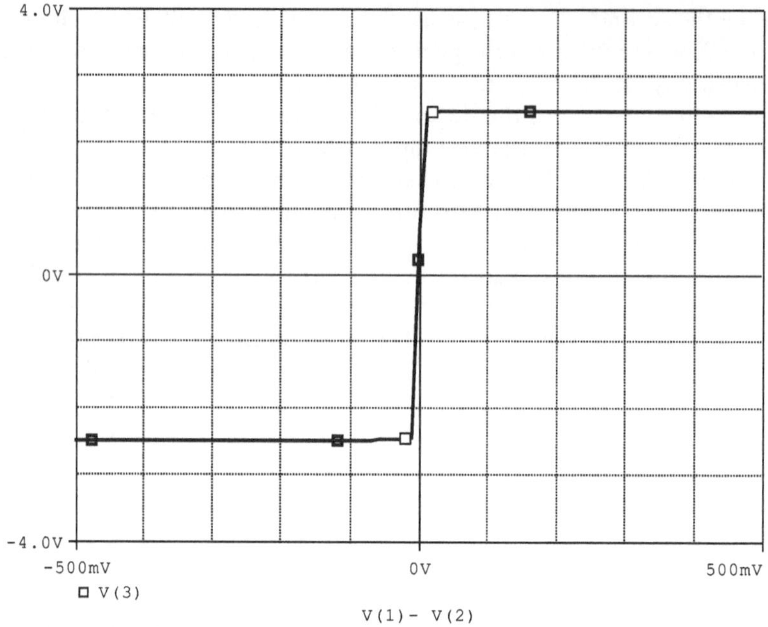

Figure 6.3 DC transfer curve showing sharp transition near zero difference input signal.

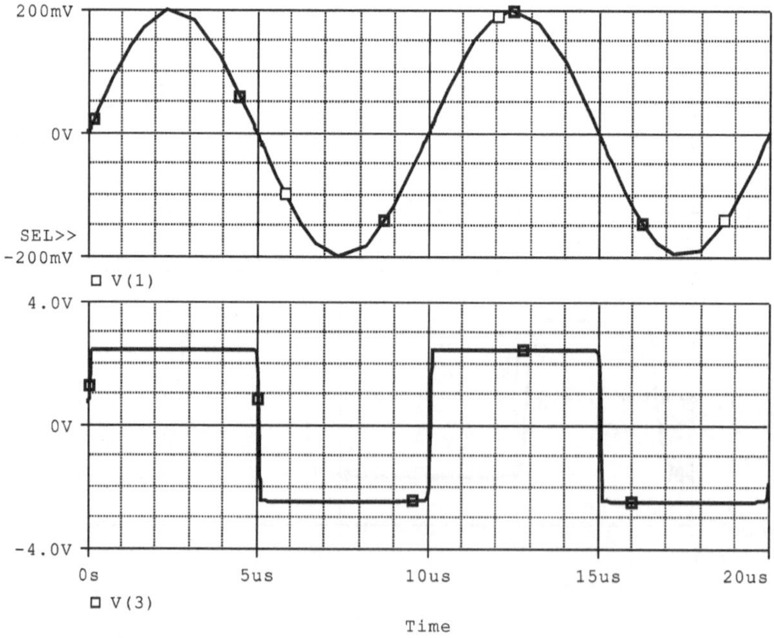

Figure 6.4 CCII application for detecting zero-crossing and working as sine to square converter.

the Y_1 and Y_2 inputs of a DVCC. The difference voltage is buffered at the X terminal. The X-terminal current is then buffered to the output Z terminal of DVCC. The advantage of using a DVCC is that both the input signals can be connected to the high input impedance terminals (Y_1 and Y_2) for voltage-mode inputs. In case of a CCII, one of the voltage inputs was applied to a low impedance (X terminal), which would cause loading of the signal. The DVCC is thus a better choice for comparing two voltage signals. The rest of the operation for comparator action is similar to the one achieved with a CCII. The difference voltage obtained at X is converted to X-terminal current, which is $(v_1 - v_2)/R_x$, and this current is available as the Z-terminal current of DVCC. The output resistance of Z terminal (high) results in the development of a voltage sufficient to saturate the DVCC, similar to the CCII case. The output voltage at Z terminal is either positive or negative saturation voltage, which depends on the supply voltage of DVCC. For the sake of gaining a better insight into the operation, the readers would appreciate the actual plots obtained using a CMOS DVCC-based design. The result of comparison of two dc signals is shown as the transfer characteristics in Figure 6.5. The DVCC is biased by a supply voltage of ±1.75V. The effectiveness of DVCC for realizing a simple comparator is thus justified. The differential signal can be applied to the DVCC input to further make this point clear. Thus, the response of the DVCC to a differential input, with X terminal grounded and Z terminal open circuited is shown in Figure 6.6. The high value of the ratio R_z/R_x ensures the saturated output, resembling a square waveform at the output for the sinusoidal differential input signal.

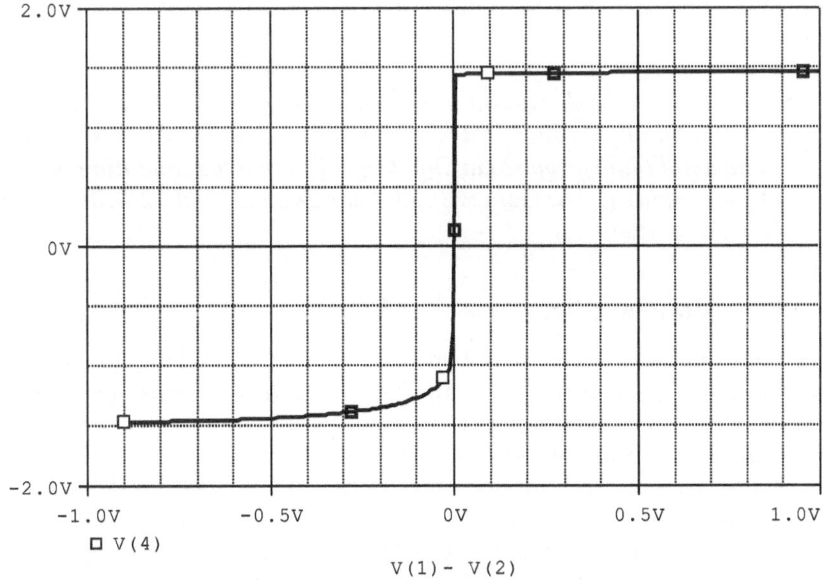

Figure 6.5 DC transfer curve for comparator designed using a DVCC.

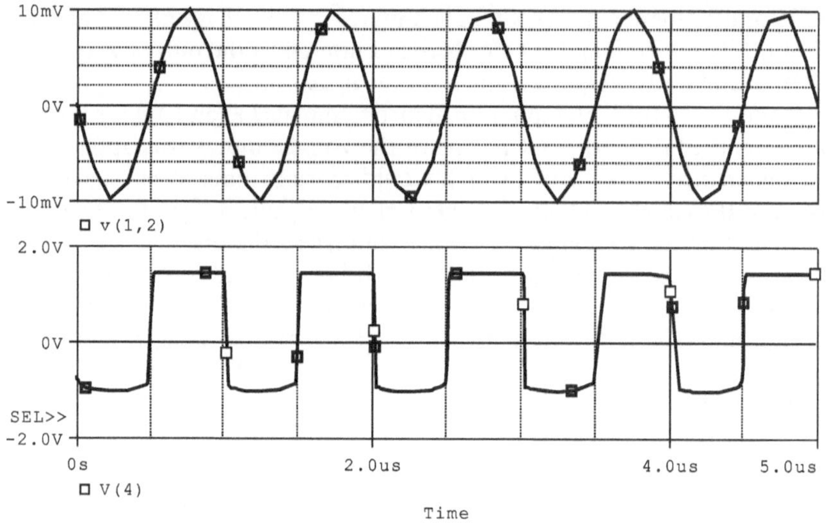

Figure 6.6 Differential sinusoidal input signal and square shape output for a DVCC.

These examples demonstrate that the current conveyors can be used to design simple comparators. The other active building blocks which are also effective in realizing comparator function include current differencing buffered amplifier (CDBA), dual X current conveyor (DXCCII), extra X current conveyor (EXCCII) and their tuneable variants. The readers are encouraged to explore the usage of these building blocks for realizing the comparator function.

Practice Exercise 6.1: *A current conveyor is biased at ±2V and has* $R_x = 100\ \Omega$ *and* $R_z = 1\ M\Omega$. *If it is used as a comparator, what is the minimum amplitude of sinusoidal input signal so that the saturation occurs at the output? (Hint: Gain is 10,000 and assume ideal ±2V saturation levels at output)*

6.2 Precision rectifiers

The next non-linear analog signal processing function is related to the conversion of an ac signal to dc signal. The change of the nature of the signal undergoing such processing is key to its classification under non-linear applications. The study of bidirectional to unidirectional signal conversion has been studied in depth using various techniques, especially the ones utilizing transformers, bridges, diodes, etc. The use of operational amplifiers for realizing rectifiers was aimed at achieving precision in signal processing. A large number of circuits for the said operation are all limited by the problems of voltage operational amplifiers. The limited slew rate and

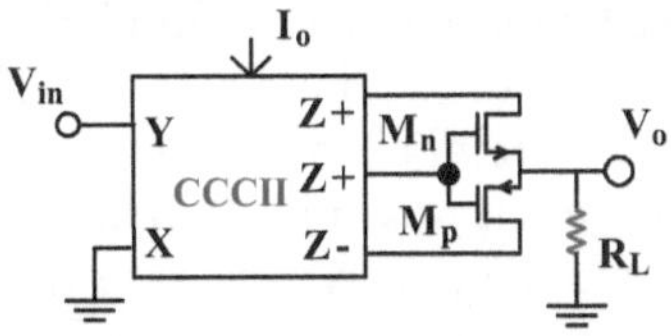

Figure 6.7 Precision rectifier circuit using a CCCII [35].

gain-bandwidth product limitations do not allow for high frequency signal processing. The processing of small signals is also restricted by the noise and offset related problems. These problems can be addressed by using current-mode techniques. The accuracy, wide bandwidth, high slew rates of current conveyor variants allow for higher performance features for designing precision rectifier circuits. One such circuit for precision rectifier is shown in Figure 6.7 [35]. It is based on a CCCII and MOS transistors. The input is connected at the high impedance node and therefore does not load the signal source. The CCCII used has multiple (three) *Z* stages.

The operation of circuit is next considered. The *Z*+ output shown in the middle is the one that is connected to the Gate terminals of two MOSFETs. This means that the impedance level at this node is high. It may be modelled simply as R_z, because the gate terminal resistance of MOS transistors is much higher than the CCCII's *Z*-terminal output resistance. The application of a small signal at the input results in a current at the *X* terminal ($i_x = V_{in}/R_x$). The *Z*+ stage output voltage is simply $(R_z/R_x)V_{in}$, which saturates because of a high ratio (R_z/R_x). It may be noted that R_x can be controlled by I_o. Depending upon the polarity of the input signal, the output at the *Z*+ terminal is either positive or negative saturation voltage ($+V_{SAT}$ or $-V_{SAT}$). The positive cycle of the input yields the positive saturation, while the negative cycle input yields the negative saturation voltage. The voltage so developed actually controls the two MOS transistors. For the positive input cycle, the M_n transistor is on, while M_p is off. For the negative input cycle, M_p is on, while M_n is off. Therefore, the positive cycle conduction at the output is through M_n and the negative cycle conduction (after inversion, due to *Z*– stage) is through M_p. The output current flows through the load in the same direction: positive cycle passing unchanged and negative cycle passing after inversion. The output is thus unidirectional. The rectification process is hence complete. The current at the drains of the two transistors can be scaled by varying the bias current of CCCII. This current is V_{in}/R_x at the drain of M_n, while it is $-V_{in}/R_x$ at the drain of M_p. The output across the load can be adjusted using I_o. This means that the output signal amplitude is electronically adjustable. Alternatively, the average value of the rectified output can be electronically controlled. This feature is unique in the sense that operational amplifier-based precision rectifiers do not possess any such facility. The circuit may be designed using either a bipolar or a CMOS implementation of CCCII. However, the use of two MOS transistors suggests that the latter design would be better. The actual simulation result obtained using the CMOS implementation is shown in Figure 6.8. The input signal of 100 mV peak amplitude and the circuit output are easy to interpret in the shown figure. The input signal frequency used is 10 kHz.

The circuit can be used as half-wave precision rectifier by removing one of the two transistors. The two circuits so realized are shown in Figure 6.9a and b. One circuit, as shown in Figure 6.9, is a positive precision rectifier. Another circuit using only a PMOS transistor is shown in Figure 6.9b. It is also a half-wave precision rectifier. It is worth mentioning that half-wave precision rectifiers find certain applications like peak detection.

The input and output of the circuit for half-wave rectification is further shown in Figure 6.10, which can be easily interpreted. For the positive input cycle, the gate voltage of M_n reaches positive saturation value (V_{SAT}), which allows for the conduction of current through the transistor. Therefore, the current flows through the transistor for positive cycle. The output voltage is available across the load resistor. For the negative cycle, the gate voltage becomes $-V_{SAT}$, thus switching off the transistor M_n. There is no path for the current to flow through the load. The output voltage remains zero. This

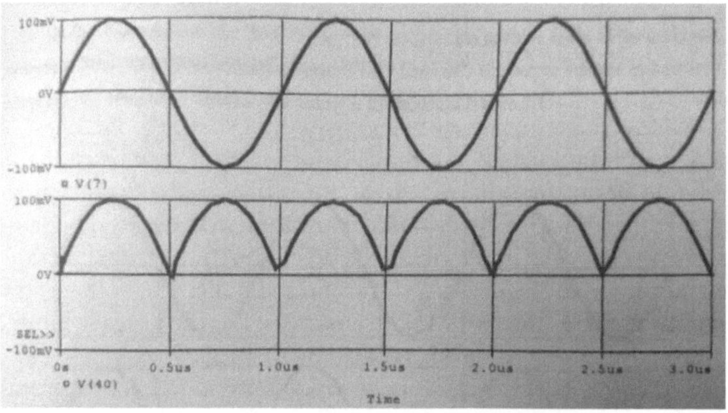

Figure 6.8 The input and output of precision rectifier circuit.

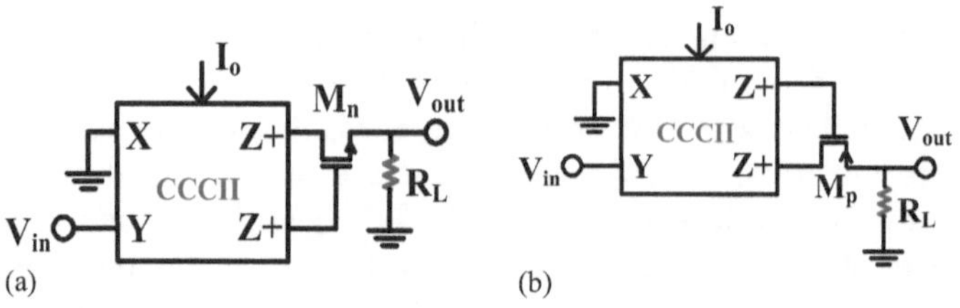

Figure 6.9 (a) Half-wave precision rectifier and (b) another half-wave precision rectifier circuit.

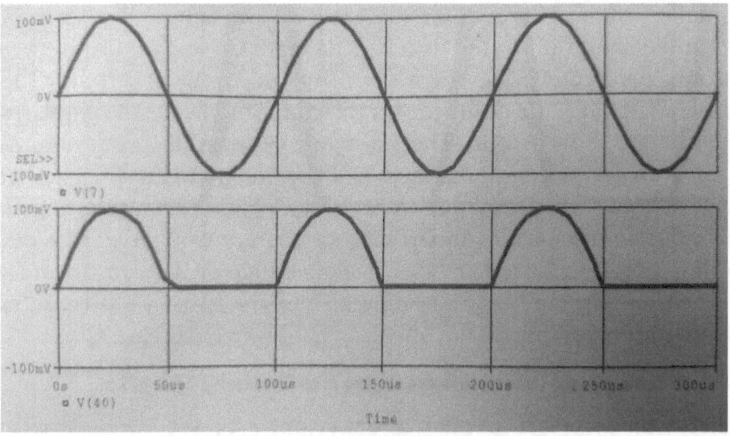

Figure 6.10 The input and half-wave rectified output.

confirms the positive half-wave precision rectification process. The readers are encouraged to analyse the negative half-wave circuit of Figure 6.9b. Is the negative cycle passed without any directional change or not? Is the positive cycle eliminated? It would be an interesting exercise worth devoting a thought to. Another exercise worth examining is the polarity of $Z+$ being changed to $Z-$, to which the M_p is connected. How is the operation of circuit changed by this change? Some of these questions are being left for interested readers.

Another circuit for precision rectifier is now presented as Figure 6.11, which requires two CCCIIs. The readers may now wonder the reason for employing an extra CCCII for the purpose. In fact, this circuit was the first one to be reported in literature, and the one presented in Figure 6.7 evolved from it.

The first CCCII in the circuit is just being used as a comparator, discussed in earlier sections. The output of the comparator acts as the control voltage for switching the two transistors. This role is fulfilled by including an

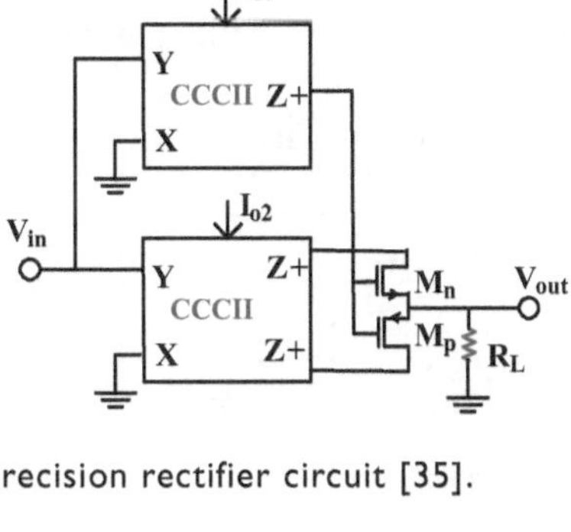

Figure 6.11 Another precision rectifier circuit [35].

additional Z stage in the single CCCII-based circuit of Figure 6.7. The CMOS implementation of the two circuits can be compared from the transistor count perspective. The single CCCII-based circuit is certainly the economical choice, since the number of transistors, and hence the chip area and power benefits, are obvious. The hardware realization using commercial chips for both the precision rectifiers is, however, no different. The two circuits can be built using three ICs in each case! This is an interesting aspect, if their integration and breadboarding options are compared. The two circuits differ in their choices: the one using a single CCCII is better targeted for fabrication, but the breadboarding cost of the two CCCII-based circuit is not different from that of the one designed using a single CCCII.

6.3 DIGITAL MODULATORS AND DETECTORS

Another very important non-linear application is required in communication which involves the transmission of digitally modulated analog signals between two or more points. The digital modulating signal is carried through the system on an analog signal, referred to as the carrier signal. Current-mode techniques offer advantageous features to perform this operation with a view to the high-speed requirements, lower supply voltage operation and the frequency of signals to be processed. The amplitude shift keying, phase-shift keying and frequency-shift keying are key digital modulation methods which find extensive applications in communication and information systems. These are referred to as ASK, PSK and FSK, respectively. Other modulation techniques are available, like BPSK (binary phase-shift keying), QPSK (quadrature phase-shift keying), MSK (minimum shift keying), etc., which are all derived from three basic keying methods. The use of current conveyors for realizing these functions is an interesting area for future communication needs. The amplitude shift keying is the simplest of all. It requires switching on and off a carrier signal in accordance to the digital modulating wave. The first circuit is given in Figure 6.12, which performs the ASK and BPSK functions.

The circuit is based on a CCII, two resistors and a MOSFET. The circuit is designed using a current follower topology, wherein the Y terminal is grounded. The circuit works on the principle of the current conveying

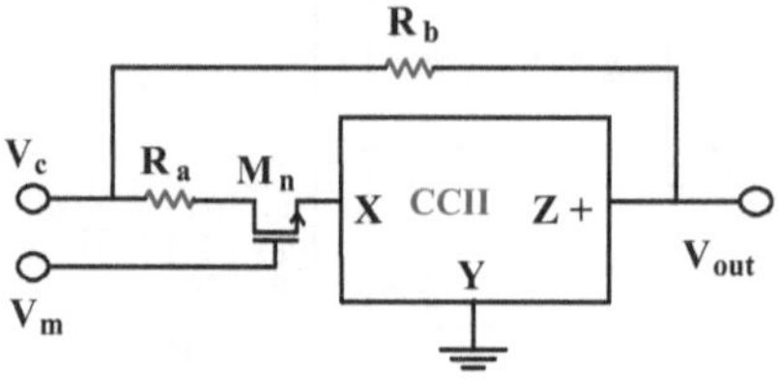

Figure 6.12 CCII-based modulator circuit.

property of current conveyors, which says that the Z-terminal current follows the X-terminal current with a good accuracy and a gain of unity. The carrier signal is marked as V_c, while the modulating signal is marked as V_m. The transistor M_n is on for a positive V_m, while off otherwise. The circuit operation begins by assuming a current through R_a as 'i_1', which is the same as the current flowing into the X terminal (i_x). The current through M_n is the same as i_x, which in turn is equal to i_x. Under the ON state of M_n, the conducting transistor offers as resistance of 'r_{on}', which is the same as the r_{ds} of the NMOS transistor. Therefore, the current $i_x = V_c/(R_a + r_{ds})$. The output current flowing into the CCII's Z terminal is i_z, which must be the same as i_x. Thus, $i_z = V_c/(R_a + r_{ds})$. But this current, as per circuit topology, is also equal to $(V_c - V_{out})/R_b$. The final expression for defining the current following property is now given below. Equating the two currents, the final expression for the output is given below.

$$\frac{V_{out}}{V_c} = \frac{R_a + r_{ds} - R_b}{R_a + r_{ds}} \tag{6.1}$$

For the close switch (M_n is ON), the ideal value of r_{ds} is assumed zero, while for the open switch, the ideal value of r_{ds} is infinite. The two values are well separated for a good switch with the appropriate design. The practical values are neither zero nor infinity in the two cases. The aspect ratio, technology used and other circuit parameters influence the exact on and off resistance values. However, ideally the following two expressions are obtained under different control voltage conditions.

(i) When V_m is high:

$$\frac{V_{out}}{V_c} = \frac{R_a - R_b}{R_a} \tag{6.2}$$

(ii) When V_m is low:

$$\frac{V_{out}}{V_c} = 1 \tag{6.3}$$

Equation (6.3) needs some justification. For the open switch, the path from the input to the X terminal is open circuit. Therefore, the output is directly connected to the input terminal through R_b, carrying no current! A unity gain is obtained between the output and the input. If $R_b = 2R_a$, then equation (6.2) results in a gain of −1. It is easy to now interpret that the circuit changes the polarity of the carrier input signal in accordance with the message signal (V_m). This property of changing the phase of carrier signal in accordance with the digital message signal realizes the phase-shift keying function. The carrier signal phase is as per the message signal. This is also referred to as binary phase-shift keying (BPSK). For message bit (0), the carrier passes without phase shift, while for message bit (1), the carrier passes after inversion. The actual CMOS-based circuit is designed and the result is given in Figure 6.13, showing the output (BPSK) along with the message signal. The circuit can be designed to perform the ASK function as well. A simple change in design is achieved if the R_b/R_a ratio is controlled. Let us keep one of the two resistors fixed and vary the other resistor. If R_a is fixed at, say, 5 kΩ, and R_b is varied, then the ASK function is obtained. Equation (6.2), which corresponds to the message bit being high, is re-written as below.

$$\frac{V_{\text{out}}}{V_c} = 1 - \frac{R_b}{R_a} \tag{6.4}$$

Now if R_b is taken as 4.5 kΩ, then the output is only 0.1 V_c! The output can thus be adjusted by varying R_b. A ratio of 0.5 is used for illustration purposes and the result is shown in Figure 6.14. The output is 0.5 times the carrier amplitude for message bit '1', while passes the carrier signal as such for the message bit '0'. This is in confirmation with the amplitude shift keying operation. It is interesting to note that the output is available at the Z terminal in the circuit. The single CCII+ circuit is easily realizable using a single AD-844 IC. The output in the IC-based realization can then be tapped from the buffered output of the chip (W), with the advantage of low output resistance. The output can thus be connected for further transmission without any loading problem.

Another circuit with a carrier signal connection at the high impedance node is discussed next. The advantage of inputting the carrier at the Y

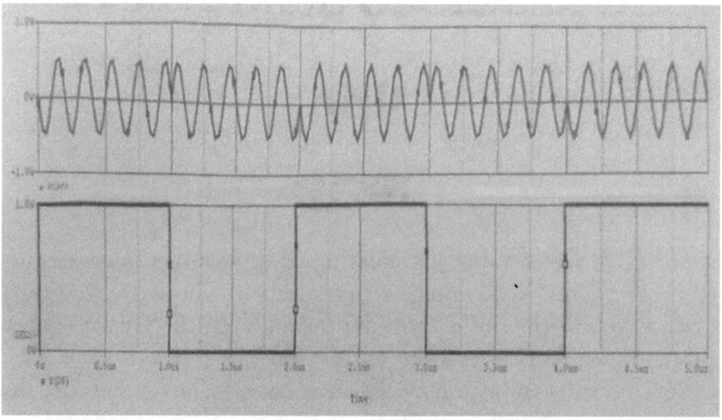

Figure 6.13 Actual result of the circuit operation as BPSK.

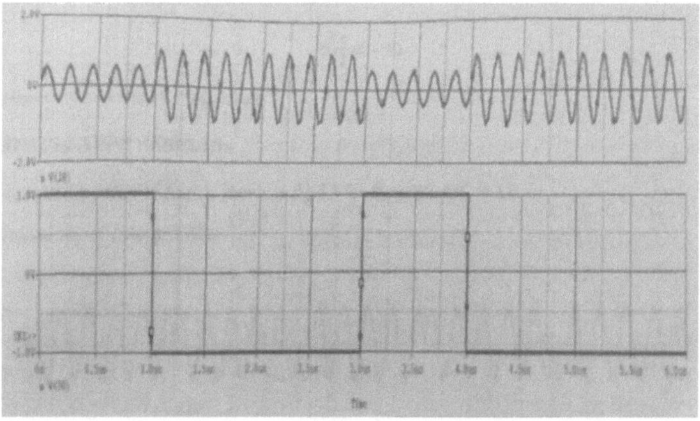

Figure 6.14 Actual result of the circuit operation as ASK.

terminal of the current conveyor is that the signal source is not loaded. The circuit is shown in Figure 6.15 and uses a single CCCII, the electronically tuneable version of CCII. The circuit does not employ a resistor in the input circuitry. The only shown resistor is in the form of load, from where the output can be tapped. The modulating signal is at the Gate terminal of MOSFET, thus there is no problem of loading the modulating signal source either. The circuit is simple and its operation easy to understand. The modulating signal would keep the transistor (M_n) on as long as it is high (digital bit remains '1'). The MOSFET is off when the modulating signal bit goes low. The X-terminal current depends on the carrier signal amplitude and the bias current. This current is available at the Z terminal. The Z-terminal current is allowed to flow across the load when the MOSFET is on, otherwise the MOSFET is like a closed switch. Therefore, the carrier signal passes the on switch when $V_m = 1$ but is blocked when $V_m = 0$. The output is accordingly available across R_L. The circuit works as an amplitude shifting keying circuit. The amplitude can be adjusted by controlling the bias current of CCCII, when the $V_m = 1$. Since the output remains zero for $V_m = 0$, the circuit is a special case of ASK, which is called on-off keying (OOK). The design of the switch is a tricky exercise, demanding a wise choice of aspect ratio, keeping in view the load. The switch resistance during on state must be low enough in comparison to the load for proper amplitude of the output

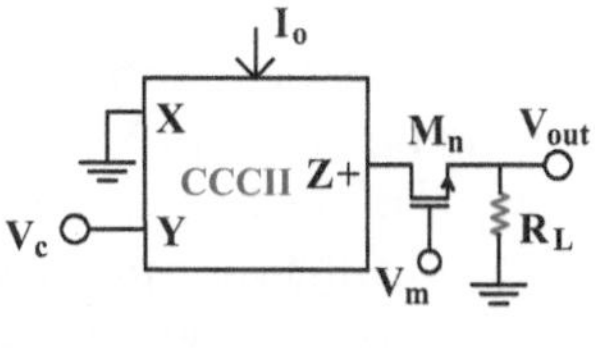

Figure 6.15 Electronically tuneable ASK circuit.

obtained. The output would be attenuated if the switch resistance is comparable with the load resistance. The design issue involves the operation of transistor in the desired region. For the ideal switching, the drain to source voltage must be zero, which means that the transistor is to operate in the triode region, where drain to source voltage is smaller than the overdrive voltage. Since the gate voltage (V_m) is digital modulating signal, it assumes binary levels (0 and 1). Logic 0 corresponds to 0 V, while logic 1 corresponds to a positive voltage, best decided by the technological constraints and the supply voltage choice. For instance, if the supply voltage for CCCII is ±2.5 V, then logic 1 may be a 2 V signal. As far as the carrier signal is concerned, it is a small signal, which needs to be modified in accordance with modulating signal. The triode region condition would also depend on the carrier signal magnitude, since the drain and hence the source voltages depend on it. If the carrier amplitude is 0.8 V, then the gate to source voltage is 1.2 V (2.0–0.8); thus the overdrive voltage for transistor threshold voltage of 0.7 V becomes 0.5 V (1.2–0.7). The transistor remains in the triode region, since the drain voltage is also equal to 0.8, and the drain to source voltage is zero! This would mean the choice of transistor aspect ratio must be made so that these voltages obey the above calculations. The only other parameters involved for the calculations are mobility and oxide capacitance for the technology being used. The transistor on resistance may be calculated from the drain to source on resistance expression, readily available in related texts. The readers are encouraged to design the circuit with the help of this example, which is only for illustrative purposes. For example, if μC_{ox} is given as 50 $\mu A/V^2$ and the switch resistance is not to exceed 1.0 kΩ, then W/L of the transistor M_n can be easily found. The aspect ratio is found as 40, using the drain to source resistance formula ($W/L = 1/\mu C_{ox} V_{ov} r_{dson}$). This example would be helpful to the readers for designing the circuit as per given specifications.

Practice Exercise 6.2: *Simulate the circuit of Figure 6.15 using bipolar implementation of CCCII. Repeat the exercise using CMOS CCCII.*

6.4 CURRENT-MODE LOGIC GATES

The use of current-mode techniques for non-linear applications is further extended to the study of another very important circuit function. This falls under the study of digital circuits and systems. The basic building block for digital circuit and system design are logic gates. The logic gates have conventionally been designed using bipolar and MOS families, with optimized transistor count and features. The features which have been considered in such designs are the power dissipation, propagation delay and fanout of realized gates. The basic motive for such designs is the integration of these circuits for very large-scale integration and design of digital systems.

The analog signal processing circuits and their merger with digital circuits for practical workable systems pose a requirement of easy compatibility. There are applications where the integration complexity is simpler and logic gates can be easily made compatible to analog circuits. The overall circuitry can then be realized using analog building blocks for compatibility reasons. This is especially feasible if the amount of digital circuitry is relatively small in comparison to the analog counterpart. Such situations are encountered in tuning of analog circuits, control of an analog parameter or simple selection of one out of many variables. The last discussed circuit in the previous section provides an excellent option for realizing logical AND operation. Let us understand the operation of the circuit in Figure 6.15 from this perspective. Assuming the two inputs marked therein, it is important to note that both these inputs are now driven by a digital signal. This means that the two input signals would assume a logic 0 or 1. When both the inputs are logic 0, the output is held to 0 V, because the transistor gate is connected to 0 V, hence it is off. If either of the input signals is logic 1, the output still remains 0 V. Two cases arise under this condition. One is when V_c is high, but V_m is low and the other when V_c is low, but V_m is high. In the former conditions, the MOS is off, hence the output remains logic 0, while in the latter condition although the MOSFET is on, it passes the logic 0, which is connected as V_c. Now the fourth case is when both the inputs are logic 1. In this case the MOSFET is on and it passes the logic 1 connected at the other input (V_c). Therefore, the circuit works as logic AND gate. The actual simulation result of the circuit is given in Figure 6.16, which justifies the use of Figure 6.15 as logical AND gate. CCII is biased at a supply voltage of ±1.5 V, while the logic 1 input voltage used is 1 V.

The realization of logic gates using current-mode techniques can be extended for other gates using various types of current conveyors. The one to be presented next in this treatment is another AND gate which provides polar outputs. The significance of polar codes is well known in communication systems. The polar AND gate circuit is shown in Figure 6.17. It is designed using a CCII, two resistors, a reference voltage and two MOS switches.

The basic topology used for the design is a current follower topology. The two resistors are to be selected with a defined ratio ($R_b/R_a = 2$). The reference voltage used in the circuit is marked as V_R, the two logic inputs are A and B, whereas the output is marked as V_{out}. The operation of the circuit is next discussed. If both the inputs are logic 0, the switches M_1 and M_2 are both off, and the output is negative of the reference voltage V_R. For the choice of V_R as a positive voltage, the output of the circuit is therefore negative voltage. This can be understood from the equivalent circuit, under the condition when two inputs are logic 0. There is no path from the V_R to the V_{out} node; hence the circuit operates an as inverter ($V_{out} = -V_R$). This is easy to obtain from the general expression for the circuit with the transfer function [$V_{out} = V_R\{1-R_b/R_a\}$]. For the resistive ratio R_b/R_a of '2', the inverting

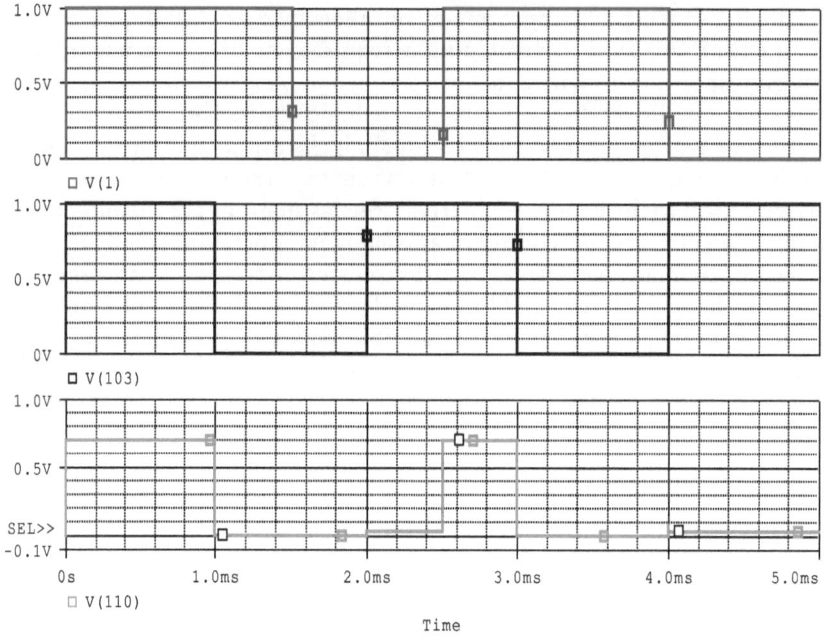

Figure 6.16 The two inputs (labelled V(1) and V(103)) and the output (V(110)).

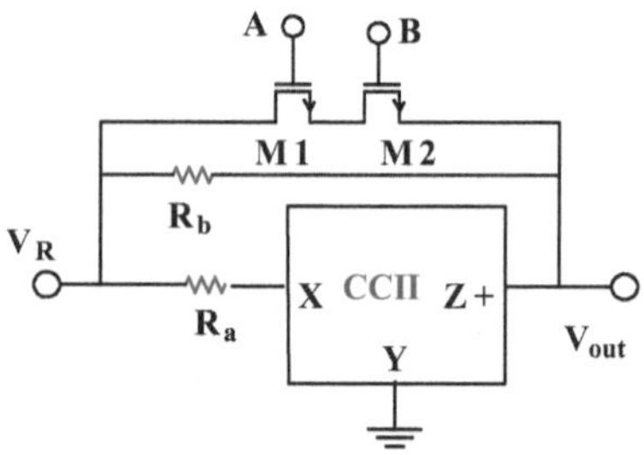

Figure 6.17 Polar logic AND gate circuit.

action of the circuit is justified. Similarly, for the two other conditions, when either of the two inputs (A or B) is logic 0, the same equivalent circuit holds good. The output thus remains $-V_R$ in these two conditions as well. The fourth condition is when both the inputs are logic 1. In this case, both the switches are on, providing a low impedance path from the V_R to the V_{out} node. Therefore, the output is simply equal to V_R, which is a positive voltage. The polar logic AND function is realized, wherein the output is positive if and only if both the inputs are logic 1, while the output is negative if any one of the inputs is logic 0. The use of 1 V reference voltage is illustrated in Figure 6.18 for presentation of actual results, wherein the top waveform corresponds to the output for the two inputs shown below the output.

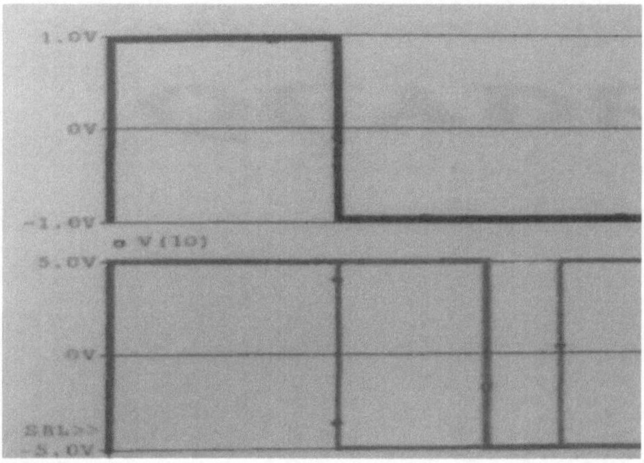

Figure 6.18 The output (above) and two inputs (below) for the polar AND gate circuit.

The inputs are also assumed to be of polar nature, assuming a positive value for logic 1 and a negative value for logic 0. The four input combinations can be easily identified in the figure and their corresponding output following polar logical AND operation.

The other polar logic gates can be similarly realized using a single CCII. For example, the logical OR function with polar output may be realized by employing MOS switches in parallel, instead of series (as in AND gate). The polar logic OR circuit is shown in Figure 6.19. In this case when either of the inputs is logic 1, there is a direct path from the V_R node to the output, thus providing a logic 1 at the output. When both the inputs are logic 0, then there is no direct path from V_R to the output. Under this condition, the circuit's inverter action yields a negative output (V_R), which corresponds to logic 0. The actual result for the circuit is shown in Figure 6.20.

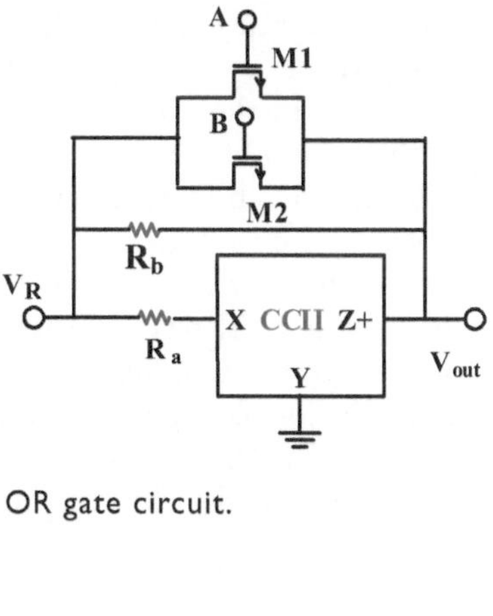

Figure 6.19 Polar logic OR gate circuit.

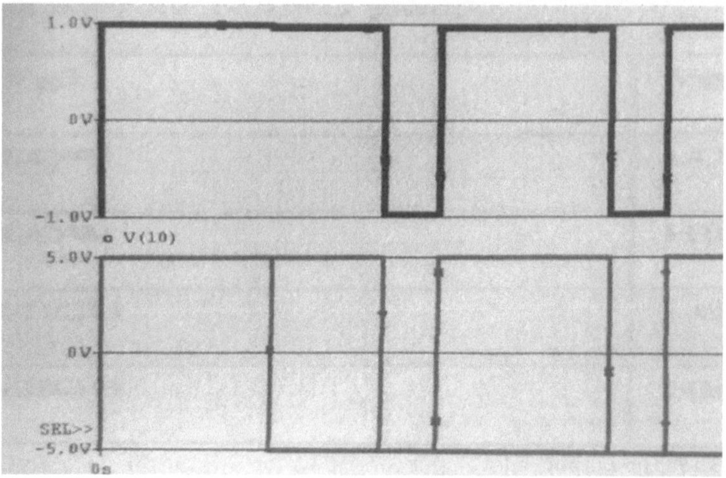

Figure 6.20 Output (above) and inputs (below) for polar OR gate.

6.5 MULTIPLIER/DIVIDER CIRCUITS

The importance of multipliers and dividers in electronic and communication systems is a well-known fact to the readers. The design of multipliers assumes importance for their useful applications in modulators, mixers, demodulators, square rooting and other non-linear signal processing functions. The analog multiplier is characterized by an output proportional to the product of two analog input signals. The design involves the use of bipolar junction transistors or, in more recent times, MOSFETs. One of the popular design methods is the use of the Gilbert cell. The operation of analog multipliers can be of different types, depending on the input and output polarities that can be processed by the circuit. Single quadrant, two quadrant and four quadrant multipliers are common, with increasing levels of functionality. The four quadrant multiplier can process signals of any polarity, with correct polarity output. The current-mode techniques for realizing multiplier circuits are beneficial for their accuracy, bandwidth and low voltage operation. One of the simplest possible current-mode techniques of realizing an analog multiplier is presented in this discussion. It is well known that logarithmic and antilogarithmic amplifiers form the basic building blocks for designing multiplier. This approach is used herein for realizing an analog multiplier. The circuit schematic is shown in Figure 6.21. It employs a CCII-based log and antilog amplifier and a summing amplifier/differencing. The basic building blocks using a CCII were introduced in literature long ago [5]. The schematic can realize a multiplier/divider depending upon the choice of the summing/differencing block. Both the inputs are applied to the high impedance nodes of the CCIIs. The resistors at the X terminals are in grounded form. Diodes are used at the Z terminals of two CCIIs driven

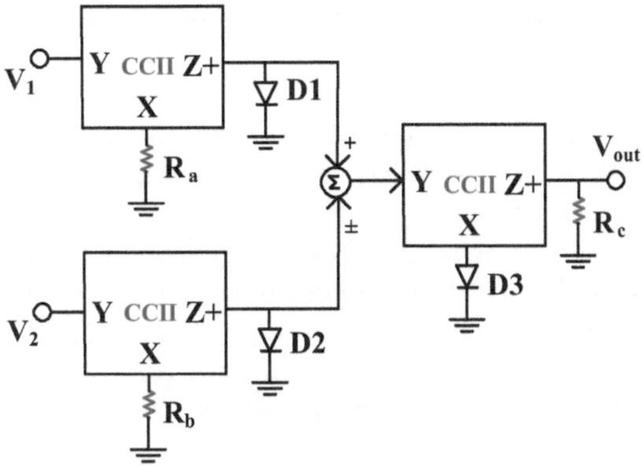

Figure 6.21 CCII+-based analog multiplier/divider circuit.

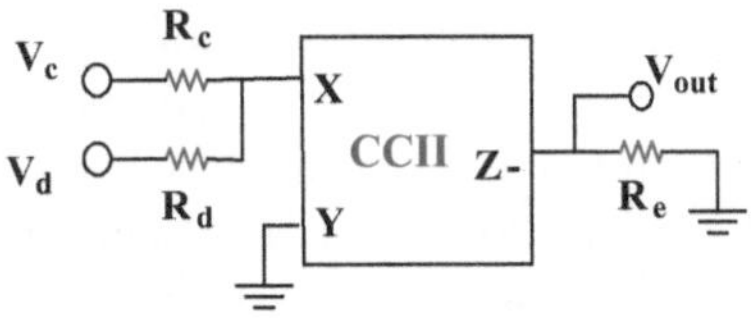

Figure 6.22 CCII-based summing amplifier circuit.

by the input signals. The first two CCIIs along with their circuitry perform the log function. The third CCII with a diode at the *X* terminal and a resistor at the *Z* terminal performs the antilog function. The summation block has an option of summing/differencing in one of its inputs. The output of the circuit is proportional to the product of two inputs, namely V_1 and V_2, for the summing block, while the output is proportional to the ratio of two inputs, namely V_1/V_2 for the differencing block. The summing block may be realized using the circuit of Figure 6.22, as one of the possible choices. Coming to the circuit of Figure 6.21, the currents flowing in the *X* terminals of the two CCIIs being driven by input signals are V_1/R_a and V_2/R_b, respectively. The *X*-terminal currents in the two CCIIs are to be conveyed to the *Z* terminals of the respective CCIIs. Hence these currents flow in the diodes D1 and D2, respectively. The standard diode current expression yields the values of these currents in terms of voltages at the *Z* terminals, which are same as diode voltages. The *Z*-terminal voltage therefore relates the input signals with logarithmic expressions. The diode voltages are now summed/differenced, before being applied to the third CCII. The third CCII's *X*-terminal current (same as diode D3 current) now depends on the output of summing/differencing block. Finally, this current flows into the resistor R_c, which allows for the equivalent output voltage. The output of summing/differencing block undergoes an antilogarithmic operation. Hence the final circuit output depends on the product/ratio of input signals. The choice of three resistors and the diodes' matching has to be appropriate for the circuit to perform satisfactorily.

Practice Exercise 6.3: *Analyse the circuit of Figure 6.21 to express the output in terms of inputs. What are the conditions for multiplication/division operation?*
Practice Exercise 6.4: *Analyse the circuit of Figure 6.22 to express the output in terms of inputs. What is the condition on the three resistor values, for the proper operation of the circuit in multiplier design?*
Practice Exercise 6.5: *For the divider circuit, the differencing block is to be designed. Show the circuit diagram of the differencing block using a single CCII.*

Another circuit for multiplier/divider can be further realized using second-generation current conveyors. The basic realization approach is similar to the previous circuit, which involves the use of log and antilog amplifier, besides a summing/differencing block. The circuit is shown in Figure 6.23.

The input signals are applied through resistors at the *X* terminals of CCIIs, which means the circuit does not enjoy high input impedance. The *Y* terminals of CCIIs are grounded. This topology is now well known to the readers, as the current follower topology. The CCIIs used are of negative polarity, unlike the previous circuit. The resistors in the circuit are in floating form, which is a drawback as compared to the circuit of Figure 6.21. The circuit component count is similar to the previous circuit. However, the CCII-based circuit of Figure 6.23 requires a greater number of ICs when compared to the previous circuit. This is an obvious drawback if the breadboarding option is to be explored. An advantage of the circuit of Figure 6.23 when compared to the earlier presented circuit (Figure 6.21) is the absence of errors due to the voltage transfer gains between the *Y* and *X* terminals. Since the *Y* terminal is grounded in current follower topology, the error in

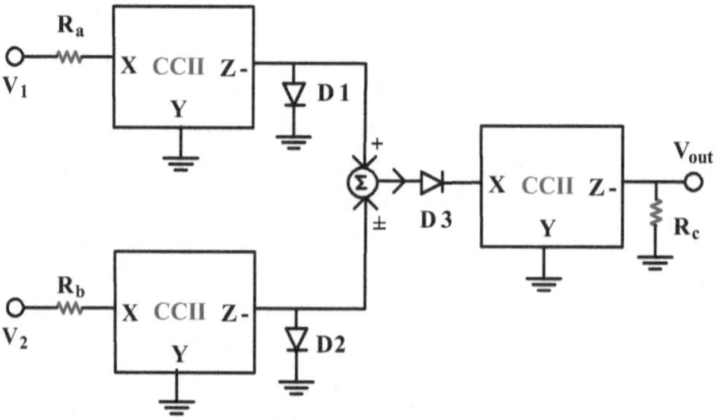

Figure 6.23 CCII−-based analog multiplier/divider circuit.

voltage transfer gain is eliminated. The only error due to CCIIs is the current transfer gain induced error. The earlier circuit (Figure 6.21) would be affected by both voltage and current transfer ratio induced errors. These errors are especially severe for high frequency operation, where the ideal unity voltage and current transfer gains start to roll-off. Moreover, these gains also become dependent on the frequency of the signals. The second circuit (current follower topology based) is advantageous in terms of higher expected frequency range and reduced errors. The non-ideal study of the circuit of Figure 6.23 can be performed. Assuming α as the current transfer gain of CCIIs, and keeping it matched for integrated conveyors, implies that the three CCIIs exhibit the same current transfer gain (α). The current through input resistors (R_a and R_b) are V_1/R_a and V_2/R_b, respectively, which flows in the Z terminals of the respective conveyors. Due to the finite current transfer gains (α in each case), these currents are $\alpha V_1/R_a$ and $\alpha V_2/R_b$, respectively. The voltages at the output of first level CCIIs are summed before being applied to the antilog amplifier circuit. The voltage of diode D3 is the sum of two voltages as above. The antilog operation is performed thereafter. The third CCII output current again introduces a factor of α. The final output expression can be found so as to study the effect of current transfer gain on the circuit performance.

Practice Exercise 6.6: *The non-ideal expression for the output of Figure 6.23 is to be analysed. Use the preceding discussion to obtain the same.*

6.6 IC COMPATIBILITY OF NON-LINEAR CIRCUITS

The circuits presented in the chapter are all compatible to be realized using commercially available ICs for current-mode operation. The use of AD-844 has already been mentioned in the earlier chapters. The circuits built using CCIIs can be experimentally tested using the IC AD-844. As a case study, the circuit of Figure 6.21 can be studied further for the purpose. The analog multiplier circuit can be built in the hardware laboratory using ICs and other elements. The other elements required are diodes and resistors. Each CCII+ requires one AD-844; thus the circuit requires three ICs for three CCIIs. The summing block is realized using Figure 6.22, which is based on a CCII–, thus requiring two AD-844 chips. The complete multiplier can thus be breadboarded using five ICs. The problem can be attempted as one of the possible projects at various levels of study programs. The readers are encouraged to test the circuit and verify its workability in real laboratory conditions. The IC can be biased by a dual dc power supply of ±10 V. However, the AD-844 can operate on a wide range of supply voltages, approximately 5–15 V, but the choice of ±10 V is quite justified as a mid-value of its range. The other circuits can be similarly tested using the available components. For example, the circuit requiring MOS transistors as switches can be built using CMOS ICs. The readers may explore the topic using recent works appearing in open literature.

6.7 SUMMARIZED CONCLUSION

This chapter explored the usefulness of current-mode techniques for realizing non-linear functions. The circuits presented in the chapter covered comparators, precision rectifiers, current-mode detectors, logic gates and various digital modulation schemes. The analog multiplier and divider using current conveyors was further discussed. Some actual simulation results were included to instigate the readers further into the subject. The variety of non-linear functions studied would broaden the scope of current-mode techniques for exploring complete system design.

Chapter 7

Waveform generation circuits

This chapter extends the study on the current-mode techniques for generation of various waveforms encountered in electronic and communication systems. The use of various current conveyors for waveform generation is explored in this chapter. The main focus is on sinusoidal oscillators with features of quadrature outputs, multiphase outputs, etc. The electronic tuning approaches of such oscillators are also discussed. The realization of oscillator circuits using different types of current conveyors is provided with critical analyses and practical considerations. The applications of oscillator circuits are discussed. The effect of non-ideal current conveyors is examined and commented upon. Illustrative design examples are included for actual simulation and experimental verification of circuits by interested readers.

7.1 TOPIC INTRODUCTION

The sinusoidal oscillators are a part and parcel of electronic and communication systems. The sinusoidal signal is one of the standard test signals for electronic and communication circuits. It is widely used in instrumentation systems as well. The sinusoidal signal with varied phase angles finds applications as carrier signal for communication. The generation of such signals is one of the most researched topics in electronic circuit design. The characteristics of circuits vary in terms of the designed parameters, which are mainly the amplitudes, frequency, phase and the realization techniques. The transistor-based circuits are designed using passive networks around bipolar junction or field effect transistors. The basic idea behind such realizations is the use of a transistor amplifier and a feedback network that fulfils the Barkhausen's criteria of oscillation. The use of transistor along with a passive tuned circuit provides realization of various types with good frequency stability. The advent of the operational amplifier led to realization techniques relying on the use of RC networks around operational amplifier(s). A large number of standard topologies are available which can be referred to as opamp-RC oscillators. The operational transconductance amplifier (OTA) paved further insights into the oscillator realization,

DOI: 10.1201/9781003403111-7

with some features which opamp-RC oscillators could not offer. The main amongst these features is the electronic tuning ability of the operational transconductance amplifier-based realizations. Both the opamp and the OTA, as also mentioned in earlier chapters, fall short of the versatility and performance features offered by current conveyors. Therefore, the current conveyor-based realizations are a favourable choice for realizing various types of oscillator circuits. The advantages are in terms of a broader range of frequency, better control over amplitude and purity of waveforms obtained in terms of reduced distortion. Moreover, the availability of both voltage and current signals is another feature which is inherent to the techniques relying on the current-mode approach. The order of oscillator network is an important parameter which has not been adequately addressed, especially using opamp and OTA-based realizations. Even through the well-known RC phase network-based designs may offer circuits which are third order, the texts available often present an impression that oscillators always imply a second order circuit. The current-mode techniques have ensured a well-defined difference in the two types of circuits: namely, the second order and the third order oscillators. The current-mode oscillators are easy to design with electronic tuning facility. The circuits based on tuneable current conveyors can offer this feature.

7.2 OSCILLATOR REALIZATION METHODS

The oscillator function in electronics and communication is characterized by two parameters of interest. These are the frequency of oscillation and the condition of oscillation. The former signifies the circuit's ability to generate a signal of particular frequency. The latter signifies the circuit's ability to generate sustained oscillations. The oscillating system of importance in signal processing is often a second order system, although third order systems are equally useful for the purpose. It is well known that any frequency selective/generating network comprises passive elements, which in our treatment would be confined to resistors and capacitors. The different realization methods thus depend on some basic networks formed by resistors and capacitors along with active building blocks. In each of the realization methods, time constants are synthesized. Therefore, the single time constant networks often become the core of many of the realization methods used to design oscillators. The earlier chapters have introduced many such networks in the form of integrators, differentiators of lossy and lossless types, all-pass filters, low-pass, high-pass and band-pass filters. Similarly, many amplifier topologies have also been introduced in earlier chapters. A number of simple interfaces have also been discussed so far in this book. All of these form the basic building blocks for realizing oscillator circuits. The common feature in all realization methods is to obtain a network with positive feedback, without any excitation input signal which generates oscillations. The condition of oscillation generation demands their amplitude to sustain a precise value, rather than dampen or grow indefinitely. Each of the possible realization methods involves a characteristic equation of second or third order. This is an important aspect of oscillator realizations. The parameters of oscillators depend on their characteristic equations. The simplest of all realization methods is the use of a lossy and a lossless integrator connected in feedback loop with the ability of oscillate. The lossy and lossless differentiators also find applications in realizing an oscillating system. The first order all-pass filter and integrator-based loop can also provide another method to realize oscillator circuits. A number of first order all-pass filters provide a method for realizing an oscillator with multiple phase outputs. A band-pass filter with passband gain is another option for realizing an oscillator circuit. The band-pass filter-based method often employs an amplifier in feedback loop, providing additional features to the realization. All of these mentioned methods are used for second order oscillators. The third order oscillator realization is possible by using second order low-pass filter and an integrator connected in closed loop. Similarly, high-pass filter-based oscillator realization is also possible with an additional single time constant network in conjunction. Several other possible realization methods involve the synthesis of a characteristic equation but are most likely to use the above-mentioned basic building blocks in their realization. The RLC resonator-based realization uses simulated active inductor and active negative resistor. This method is also very popular for oscillator design. The use of well-known bridge networks is another possible choice for realizing oscillator circuits. With the background on different available realization methods, the following sections would now confine to the use of current-mode techniques for the purpose.

7.3 QUADRATURE OSCILLATOR CIRCUITS

The generation of sinusoidal signals with precise phase shift of 90° is of significance in numerous communication and instrumentation systems. The circuits which provide such signals are referred to as quadrature oscillators. The quadrature oscillators with two or more outputs would generate sinusoidal signals with successive 90° phase shifts. The circuits providing two, three and four outputs are normally easily realizable. The generation of voltage, current or both voltage as well as current outputs is also possible. The use of CCIIs and passive components in the form of resistors and capacitors can be one of the current-mode approaches for realizing an active-RC current-mode quadrature oscillator. The first such oscillator with quadrature outputs is presented in Figure 7.1, which is based on CCIIs and passive elements. It requires two CCIIs, three resistors and two capacitors. The circuit is the CCII-RC equivalent of the tuneable network of ref. [16]. It is based on one of the realization methods discussed in the preceding section

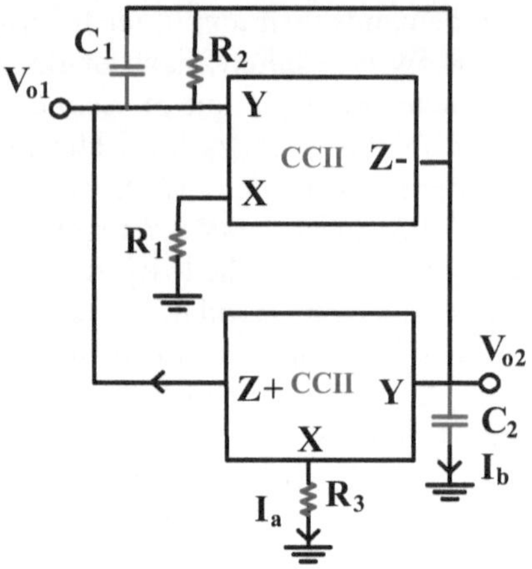

Figure 7.1 Quadrature oscillator circuit using two CCIIs.

based on the use of an all-pass filter and integrator sections in a closed loop connection.

The current-mode all-pass filter and lossless integrator are the basic building blocks used for the realization. These two current-mode blocks are connected in cascade with a closed feedback loop. The CCII-1, R_1, R_2 and C_1 form the first order current-mode all-pass filter. The CCII-2, R_3 and C_2 form the current-mode lossless integrator circuit. The output of the all-pass filter is fed to the integrator circuit, whose output is fed back to the input node of the all-pass filter. This method is one of the popular methods for realizing a quadrature oscillator. The second order characteristic equation is obtained by using the standard CCII-defining equations. A step-by-step analysis is given below.

$$\left(v_{01} - v_{02}\right)\left(sC_1 + \frac{1}{R_2}\right) = \frac{v_{01}}{R_1} + v_{02}sC_2 \tag{7.1}$$

$$\left(v_{01} - v_{02}\right)\left(sC_1 + \frac{1}{R_2}\right) = \frac{v_{02}}{R_3} \tag{7.2}$$

The above equations (7.1 and 7.2) need to be solved for obtaining the oscillator characteristic equation. For the sake of simplicity, some assumptions are now made. Let us use a design with two equal resistors, namely R_1 and R_2, so the two resistors may be referred to as 'R'. Similarly, let us assume two equal-valued capacitors, so each capacitor then becomes 'C'. The resulting characteristic equation is expressed as below.

$$s^2 + s\left(\frac{2}{RC} - \frac{1}{R_3C}\right) + \frac{1}{R^2C^2} = 0 \tag{7.3}$$

The solution of equation (7.3) requires replacing 's' with '$j\omega$', then equating the real and imaginary components on the two sides of the equation. The following frequency of oscillation (FO) and condition of oscillation (CO) are obtained. The FO is represented as ω_0, while the CO shows a sign of inequality. This is as a result of the oscillator circuit's condition of sustained oscillations. The location of poles is slightly oriented towards the right half of the s-plane for the oscillations to start. As soon as the oscillations grow, the circuit's non-ideality would bring back the poles on imaginary axes. The Barkhausen's criterion of oscillations also demands such an inequality in terms of sustained oscillations. Therefore, the negative component in the multiple factor of the s-term in equation (7.3) has to be slightly more than the positive component. The resulting condition as obtained from equation (7.4) requires the resistor R_3 to be slightly smaller than half of the other resistors (R).

$$\omega_0 = \frac{1}{RC}; R_3 \leq \frac{R}{2} \tag{7.4}$$

Another important feature of the circuit is the independent control over the frequency of oscillation by two equal-valued resistors (R). However, the condition of oscillation then requires to be set such that it is around 0.5R. The outputs obtained from the circuit are marked as v_{o1} and v_{o2} for the voltage outputs and I_a and I_b for the current outputs. The voltage outputs are phase shifted as per equations 7.1 and 7.2. However, the current output relation is given as below.

$$I_b = sR_3C_2I_a \tag{7.5}$$

Equation (7.5) suggests that the two current outputs are related with quadrature property. Replacing 's' with '$j\omega$', and at the frequency of oscillation 'ω' with 'ω_o', the following expression is obtained:

$$I_b = j\omega_o R_3 C_2 I_a \tag{7.6}$$

For equal capacitor design, $C_2 = C$, and using the FO expression (7.4), the current relationship of equation (7.6) is simplified as below.

$$I_b = j\left(\frac{R_3}{R}\right)I_a \tag{7.7}$$

The CO demands that R_3 is half of R; therefore equation (7.7) is simplified as below.

$$I_b = j\left(\frac{1}{2}\right)I_a \tag{7.8}$$

Equation (7.8) suggests that I_b leads I_a by 90°, and the amplitude of I_b is half of I_a. Thus, the circuit provides quadrature current outputs. The two current outputs are available either across a capacitor or through a resistor. This is a problem as far as sensing the outputs is concerned. The circuit requires additional current followers to sense the current outputs at high impedance nodes. The readers are encouraged to use two current followers at the marked current outputs in Figure 7.1.

Practice Exercise 7.1: *Design the quadrature oscillator circuit of Figure 7.1 for a frequency of 1 MHz using capacitors of 50 pF.*

Practice Exercise 7.2: *Find the phase relationship for the two voltage outputs in oscillator circuit of Figure 7.1.*

Practice Exercise 7.3: *The quadrature oscillator of Figure 7.1 is designed using 40 pF capacitors. The two resistors (R_1 and R_2) are taken as 4 kΩ each. Find the value of R_3. Also calculate the FO.*

Practice Exercise 7.4: *Use CMOS implementation of CCII (given in earlier chapters) to simulate the oscillator of Figure 7.1 using the available technology node parameters.*

The next quadrature oscillator circuit using current-mode techniques to be presented is based on DVCCs and passive elements. The circuit requires two DVCCs, two resistors and two capacitors. The circuit is shown in Figure 7.2 [36].

The DVCC used in the circuit employs multiple Z stages. All the four passive elements are in grounded form. Moreover, the passive elements' count is minimum for realizing an active-RC oscillator of second order. The circuit can be analysed using the DVCC-defining equations, already presented in earlier chapters. The characteristic equation of the oscillator is given as below.

$$s^2 + s\left(\frac{1}{R_1C_1} - \frac{1}{R_2C_2}\right) + \frac{1}{R_1R_2C_1C_2} = 0 \tag{7.9}$$

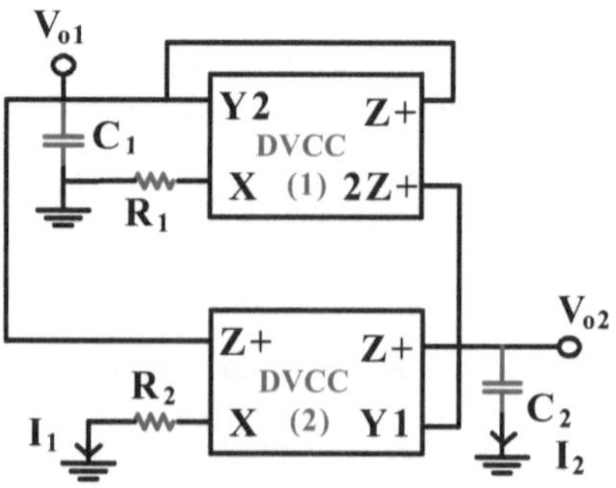

Figure 7.2 Quadrature oscillator circuit using two DVCCs [36].

The FO and CO can be obtained from equation (7.9) in a similar manner as obtained from the earlier presented circuit. The following expression gives the oscillator's FO and CO.

$$\omega_o = \sqrt{\frac{1}{R_1R_2C_1C_2}}; R_2C_2 \le R_1C_1 \tag{7.10}$$

Equation (7.10) suggests that both the FO and CO depend on the four passive elements, which means that independent control over oscillator frequency and condition is not possible. It may be noted that the circuit shows both voltage and current outputs. The two voltage outputs and two current outputs are provided by the circuit of Figure 7.2. The relationship between the two current outputs is given below.

$$I_2 = sR_2C_2I_1 \tag{7.11}$$

At the frequency of oscillation, equation (7.11) is simplified as below.

$$I_2 = j\omega_oR_2C_2I_1 \tag{7.12}$$

From equation (7.12), it is evident that the two currents exhibit a quadrature relationship. For an equal resistors and capacitors-based design, the amplitudes of the two outputs are also equal. The two voltage outputs are related as below.

$$v_{o2} = \left(\frac{R_2}{R_1} + sR_2C_1\right)v_{o1} \tag{7.13}$$

The phase shift between the two voltage outputs can be easily found from equation (7.13). One question arises regarding the quadrature oscillator circuit of Figure 7.2. Are the current outputs easy to measure? The answer is obviously no! It is due to the fact that the current outputs are across the passive elements. In order to practically measure these outputs, additional current sensing elements in the form of current followers are required. Another question which is quite obvious is regarding the tuning of the circuit. Is it possible to vary the FO without disturbing the condition, and vice versa? Again, the readers must note the involvement of four passive elements in both the FO and the CO expressions. This means that independent control over FO and CO is not possible. Moreover, is it fair to expect independent control over FO and CO in a circuit with four passive elements realizing a second order oscillator? The answer to this question is non-affirmative again. Some of these aspects are worth pondering for the need of circuits with independent control over FO and CO.

Practice Exercise 7.5: *The circuit of Figure 7.2 is to be designed for FO = 2 MHz. If the capacitors are to be 100 pF, find the value of two resistors.*

The quadrature oscillator circuit of Figure 7.1 is still a good choice despite the above-mentioned limitations. It is good from the perspective of parasitic absorption. Let us next study the circuit from the parasitic absorption viewpoint. Looking into the circuit topology, it is clear that resistors are employed at the *X* terminals of each of the two DVCCs. Moreover, in both of the DVCCs, the *Y* and *Z* terminals are shorted, with capacitive terminations in each case. As already mentioned earlier, the four passive elements are in grounded form. These three features combined together give the circuit its ability to absorb parasitic effects. It is well known to the readers at this stage that current conveyor-based circuits with a resistive terminal at *X* are good at absorbing the intrinsic *X*-terminal resistance of current conveyor (Rx). Similarly, the shorted *Y* and *Z* terminals of current conveyors allow the parasitic capacitances at these nodes to merge (add) with external capacitors at these nodes; the parasitic capacitances (of the *Y* and *Z* terminals) are also absorbed. How does this topology actually work in reality? The two external resistors are slightly increased due to the *X*-terminal resistance of current conveyors. The two external capacitors are also slightly increased as a result of the parasitic capacitances at the *Y* and *Z* nodes. The effective FO and CO can therefore be expressed as below.

$$\omega_o' = \sqrt{\frac{1}{R_1'R_2'C_1'C_2'}};R_2'C_2' \leq R_1'C_1' \tag{7.14}$$

In equation (7.14), the changed values of components are expressed as below.

$$R_1' = R_1 + R_{x1}; R_2' = R_2 + R_{x2}; C_1' = C_1 + C_{P1}; C_2' = C_2 + C_{p2} \tag{7.15}$$

Equation (7.15) expresses the effective components, including parasitic values. The two parasitic capacitances with suffix '*p*' refer to the *Y-Z* shorted nodes of the two DVCC, and comprise the parasitic capacitance at the *Y* and *Z* terminals in each case. It seems that the non-ideal expression of FO (equation 7.14) is quite different from the ideal value. The four elements are increased due to parasitic effects. As a result, the actual FO obtained would be slightly less than the designed value. The circuit can be designed with predistorted components to minimize the effect of parasitic elements.

Practice Exercise 7.6: *The oscillator circuit in Figure 7.2 is designed using resistors of 2 kΩ and capacitors of 100 pF. The X-terminal resistance of DVCC is 80 Ω. The Y- and Z-terminal capacitances of DVCC are 0.1 pF. Calculate the discrepancy between the designed and the actual FO.*

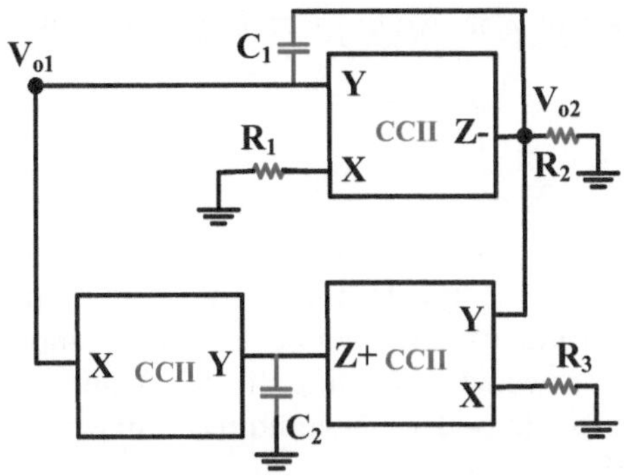

Figure 7.3 Quadrature oscillator circuit with independent frequency control.

Another active-RC quadrature oscillator circuit is next described using CCIIs and passive components. The circuit with three CCIIs, three resistors and two capacitors is shown in Figure 7.3.

The circuit provides quadrature voltage outputs. The core of this circuit is an all-pass section, integrator and a voltage follower. The CCII-1 along with passive components C_1, R_1 and R_2 forms the first order all-pass filter [15]. The CCII-2 along with C_2 and R_3 forms the integrator block. The CCII-3 is being used as a voltage follower. The circuit topology realizes a second order quadrature oscillator with the following characteristic equation.

$$s^2 + s\left(\frac{1}{R_2C_1} - \frac{1}{R_3C_2}\right) + \frac{1}{R_1R_3C_1C_2} = 0 \tag{7.16}$$

From equation (7.16), the FO and CO can be obtained as below.

$$\omega_o = \sqrt{\frac{1}{R_1R_3C_1C_2}}; R_2C_1 \le R_3C_2 \tag{7.17}$$

Equation (7.17) can be interpreted for the variation of FO and setting of condition of oscillation. It is evident that the CO expression does not involve R_1. This provides an easy opportunity to vary the FO without disturbing the CO. Therefore, the circuit provides independent control over FO and CO. The CO can be set through either of the two remaining resistors, namely R_2 or R_3. However, it is desirable to set the condition through R_2, since the FO expression does not involve R_2. These features make the circuit especially attractive from the non-interactive tuning perspective. Non-interactive tuning implies control over FO and CO by independent elements. In this case, these independent elements are R_1 and R_2 for tuning the FO and CO, respectively. Another feature of the circuit is the use of grounded components, except for one capacitor. The two outputs provided by the circuit are related as below.

$$v_{o2} = \left(sR_3C_2\right)v_{o1} \tag{7.18}$$

From equation (7.18), it is evident that the outputs are in quadrature, with v_{o2} leading v_{o1} by 90°. At the frequency of oscillation, the relationship between the two outputs can be obtained for their amplitudes. The phase relation does not get affected by varying the FO. A simple design with equal resistors and capacitors in each case would result in equal amplitudes of the two outputs. The IC compatibility aspect of the circuit can be further studied. The circuit employs a CCII– and two CCII+. One of the two CCII+ is being used as a voltage follower. The CCII– requires two AD844 ICs, while the CCII+, being used for integration function requires another IC. The CCII being used as a voltage follower does not require additional IC because the integrator output can be tapped from the *W* terminal of the IC being used for the purpose. Therefore, the quadrature oscillator circuit of Figure 7.3 can be breadboarded using only three ICs. It may be recalled that the circuit of Figure 7.1 presented earlier would also use three ICs for experimental breadboarding. This aspect is of special interest to the hobbyist at various levels of study. This aspect is of equal significance to instructors/teachers setting up laboratory experiments on analog signal processing. These circuits can also be pursued as mini-projects by undergraduate electronics engineering students. The design of current conveyors using CMOS technology under simulation environment and their subsequent application in implementing the quadrature oscillator discussed herein is another area of study. The complete layout of the circuit using available tools is a problem relevant for undergraduate and Master's level projects. The final chip design of the circuits and their specific application is another intricate problem of attraction for chip designers. There is a lot of scope for interaction between industries and institutes to design actual products based on some of these circuits.

Practice Exercise 7.7: *Design the quadrature oscillator of Figure 7.3 for FO = 10 MHz using 50 pF capacitors.*

7.4 MULTIPHASE OSCILLATORS

The difference between quadrature oscillators and multiphase oscillators is to be first noted before actually going into this section. The quadrature oscillators are supposed to provide 90° phase-shifted waveforms. However, multiphase oscillators are supposed to provide waveforms which are shifted in phase by any other angle. The phase difference between two outputs in multiphase oscillator can be 30°, 45°, etc. Henceforth, the multiphase oscillators would mean to imply the oscillator circuit providing phase shifts other

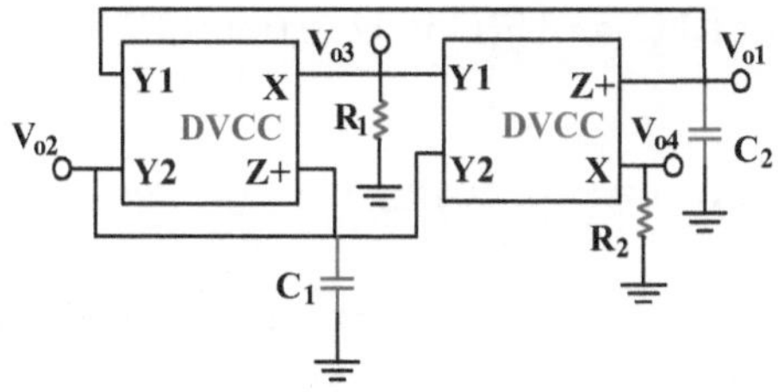

Figure 7.4 Multiphase oscillator with $\pi/4$ shifted outputs [37].

than 90°. This section is therefore confined to such circuits designed using current-mode techniques. The first circuit to be presented under this category is designed using DVCCs and passive elements. The circuit is shown in Figure 7.4 [37].

It requires two DVCCs and four passive elements, a count which is minimum for a second order active-RC oscillator. The circuit employs all four passive elements in grounded form. It provides four voltage outputs, which are progressively phase shifted by 45°. The characteristic equation for the circuit is given as below.

$$s^2 + s\left(\frac{1}{R_1C_1} - \frac{1}{R_2C_2}\right) + \frac{1}{R_1R_2C_1C_2} = 0 \tag{7.19}$$

From equation (7.19), the FO and CO is found as below.

$$\omega_o = \sqrt{\frac{1}{R_1R_2C_1C_2}};R_2C_2 \leq R_1C_1 \tag{7.20}$$

Equation (7.20) suggests that the FO and CO control is possible by four passive elements. The four outputs are related as below.

$$v_{o4} = (sR_2C_2)v_{o1};v_{o3} = (sR_1C_1)v_{o2};v_{o1} = (1 + sR_1C_1)v_{o2} \tag{7.21}$$

Equation (7.21) needs to be interpreted for the phase and magnitude relationship of the four outputs. The output v_{o4} leads v_{o1} by 90°, while the output v_{o3} leads v_{o2} by 90°. The output v_{o1} leads v_{o2} by 45°, which in turn implies that the four outputs are separated in phase by 45° with progressive leading order of v_{o2}, v_{o1}, v_{o3} and v_{o4}. Let us consider equal component design, such that the resistors used are of value R each, and capacitors used are of value C each. Then at the frequency of oscillation, the simplified relationship is obtained as below. It may be noted that $s = j\omega$ substitution has also been made.

$$v_{o4} = (j\omega RC)v_{o1};v_{o3} = (j\omega RC)v_{o2};v_{o1} = (1 + j\omega RC)v_{o2} \tag{7.22}$$

The substitution of FO as $1/RC$ in equation (7.22) yields the following relationship.

$$v_{o4} = \left(j\omega / \omega_o\right)v_{o1}; v_{o3} = \left(j\omega / \omega_o\right)v_{o2}; v_{o1} = \left(1 + j\omega / \omega_o\right)v_{o2} \tag{7.23}$$

From equation (7.23), it is clear that the magnitudes of v_{o4} and v_{o1} at the frequency of oscillation are the same, whereas the magnitudes of v_{o3} and v_{o2} are the same. However, the output v_{o1} magnitude is 1.414 (square root of 2) times the output v_{o2}. Coming to the parasitic consideration of the oscillator circuit of Figure 7.4, it is interesting to note that the grounded components are not only favourable from the integration perspective but also absorb DVCC parasitic elements. The resistors are connected at the *X* terminals, while capacitors are connected at *Y-Z* shorted terminals of DVCCs. The *X*-terminal intrinsic resistance of DVCCs are easily merged with external resistors. The *Y-Z* terminals' parasitic capacitances merge with external capacitors. The effective values of chosen external components are slightly enhanced, but there is a scope for easy correction in oscillator parameters. From the estimation of DVCC parasitic resistances and capacitances, which is often done using advanced simulators, the external components can be judiciously chosen to compensate for the parasitic-induced deviations. Therefore, it is possible to obtain accurate frequency of oscillation by appropriately designing the circuit.

The multiphase oscillator circuit is next designed using equal capacitors (10 pF) and equal resistors (4 kΩ) for the oscillation frequency of approximately 4 MHz. The DVCC is implemented using the CMOS circuitry presented in earlier chapters. The actual simulation results are presented to show the utility of the circuit as multiphase oscillator. The four outputs are plotted in Figure 7.5, where a progressive phase shift of 45° is quite evident. The readers are encouraged to work out the phase relationship from Figure 7.5 and verify the same using the expression given in equation (7.23). The four outputs are not of equal amplitudes as already mentioned earlier. The four outputs can be obtained with equal amplitudes by augmenting the circuit with an additional scaling network. The outputs with lower amplitudes can be amplified using scale changing circuit. Such a circuit can be easily design using current conveyors. For the present case, it is best advisable to use DVCC itself for modularity reasons. Therefore, two additional DVCC and resistors'-based circuits can be employed for realizing equal outputs. The readers are encouraged to design such scale changing circuits. However, for the readers to appreciate this aspect, another set of results are given in Figure 7.6. The modified circuit with additional scale changing network is simulated to obtain the equal amplitude outputs. The circuit with equal outputs is of special interest for $\pi/4$ quadrature phase-shift keying applications. The circuit can be further modified to generate eight outputs. Each of the four outputs obtained from the circuit can be inverted using inverter

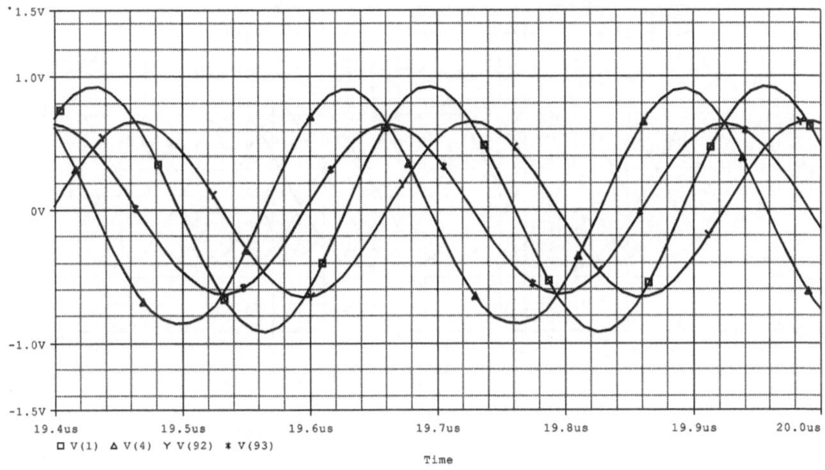

Figure 7.5 The four outputs of the multiphase oscillator circuit.

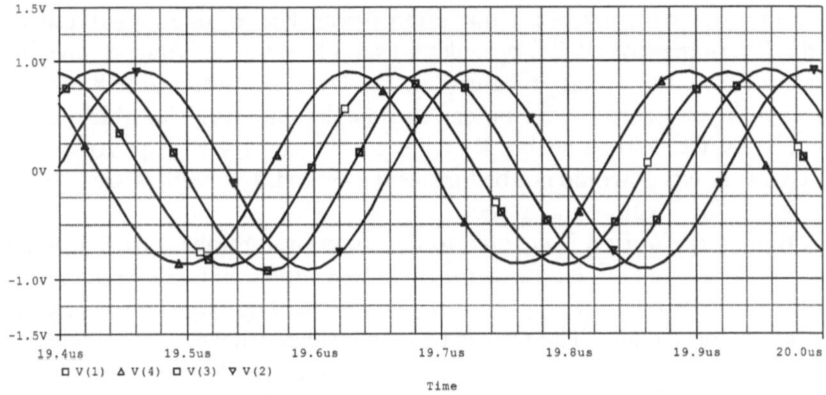

Figure 7.6 The multiphase oscillator circuit with equal outputs.

circuits. The inverter circuit can be easily designed using DVCC itself. Thus, the modified circuit would provide eight outputs, with a progressive 45° phase shift. Such a circuit with eight phase outputs finds useful applications in communication systems. The actual simulation results are shown in Figure 7.7, which presents the eight outputs.

Practice Exercise 7.8: *Design the multiphase oscillator circuit of Figure 7.4 using 20 pF capacitors for an oscillation frequency of 8 MHz.*

Practice Exercise 7.9: *The circuit of multiphase oscillator given in Figure 7.4 is to be designed using 20 pF capacitors. If the DVCC exhibits a parasitic capacitance of 0.15 pF each at Y and Z terminals, what is the actual FO obtained?*

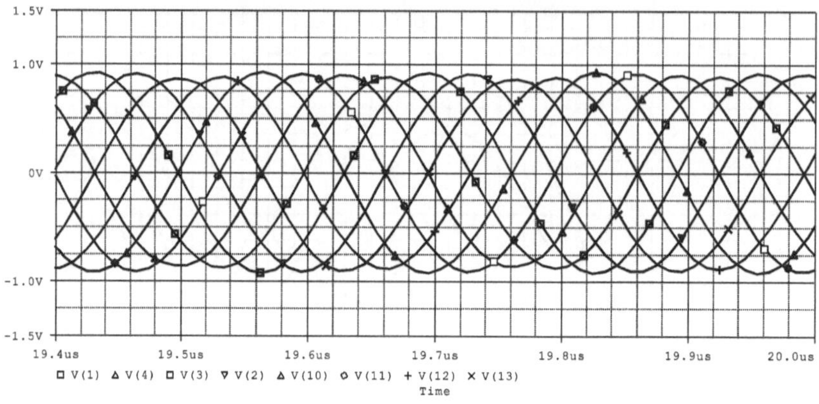

Figure 7.7 The eight outputs of multiphase oscillator circuit.

The next aspect of interest to be shared with readers is the tuning possibility of the circuit of Figure 7.4. As already mentioned, the circuit is quite favourable from the parasitic elimination aspect. The use of external resistor at the X terminals of DVCC provides a useful hint for designing a tuneable oscillator from the circuit of Figure 7.4. The DVCC with tuneable R_x can be used for the purpose. The circuit can then be designed as a tuneable oscillator. The DVCC with tuneable R_x relies on the control of bias voltage of DVCC. This aspect would not be further elaborated. The readers are encouraged to further explore this aspect.

Although the section is devoted to the study of multiphase oscillators, it may be noted that oscillators with multiple outputs with different phase shifts fall under this category. The quadrature oscillator section has already covered some circuits. The circuit to be considered next may be called a multiphase oscillator, but actually provides several quadrature outputs. It is based on a band-pass filter with passband gain, which when connected in feedback provides oscillations. The circuit is based on DVCCs and passive elements. The circuit is shown in Figure 7.8 [38].

The circuit employs three DVCCs and one DDCC (differential difference current conveyor). It is important to point out that the DDCC is being referred to for the first time in this text and needs some introduction. It is basically similar to a DVCC, but with an additional input terminal (Y_3). It employs Y_1, Y_2 and Y_3, all three as high impedance voltage inputs. The X-terminal voltage is $v_x = v_{y1} - v_{y2} + v_{y3}$. The rest of the characteristics of the DDCC are similar to the DVCC. However, in the circuit of Figure 7.8 the DDCC used does not employ the Z stage. It is simply used for summing/differencing of signals. The DVCCs used in the circuit employ Z+ stages in each case. The two capacitors and one resistor in the circuit are grounded, whereas the three resistors are in floating form. Now the circuit can be analysed for its characteristic equation.

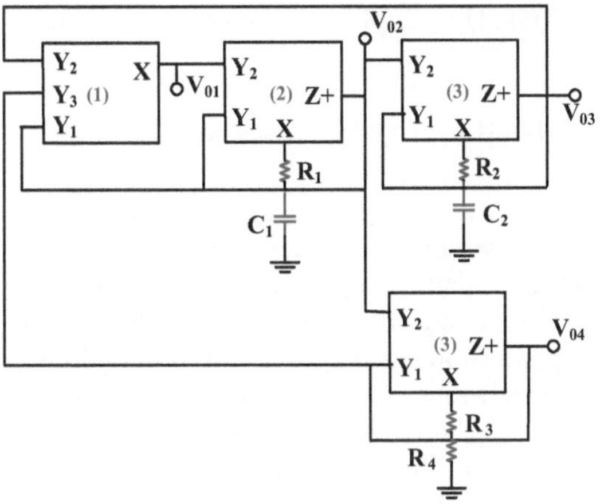

Figure 7.8 Multiphase sinusoidal oscillator circuit using DVCCs [38].

$$s^2 + s\left(\frac{2}{R_1C_1} - \frac{4R_4}{R_1R_3C_1}\right) + \frac{4}{R_1R_2C_1C_2} = 0 \quad (7.24)$$

The FO and CO can be obtained from equation (7.24) as below.

$$\omega_o = \frac{2}{\sqrt{R_1R_2C_1C_2}};2R_4 \geq R_3 \quad (7.25)$$

Equation (7.25) is now interpreted for the possibility of controlling FO and CO. Since the FO does not depend on R_3 and R_4, the CO is easy to set through R_3 and/or R_4. The FO depends only on R_1 and R_2; therefore independent control of FO is possible through these two resistors. The circuit can be tuned such that the FO and CO do not depend on each other. This is an important property for an oscillator and is called non-interactive control over oscillation frequency and condition. The four marked outputs of the circuit are now expressed for their relationship as below.

$$v_{o2} = -\left(\frac{sR_2C_2}{2}\right)v_{o3};v_{o1} = -\left(\frac{sR_1C_1}{2}\right)v_{o2};v_{o4} = -v_{o2} \quad (7.26)$$

Equation (7.26) suggests quadrature relation between v_{o2} and v_{o3}, as well as between v_{o1} and v_{o2}. This implies a progressive 90° phase shift in v_{o3}, v_{o2} and v_{o1}. The fourth output (v_{o4}) is inverted with respect to v_{o2}. Thus, it leads v_{o1} by 90°, completing the progressive phase shift order. One simplification in terms of design can be made for further understanding the interrelationship between the four outputs. Let us consider an equal resistors ($R_1 = R_2 = R$) and equal capacitors ($C_1 = C_2 = C$) based design. The FO for this design is $\omega_o = (2/RC)$. The simplified relation for the four outputs is then obtained as below.

$$v_{o2} = -\left(j\omega / \omega_o\right)v_{o3}; v_{o1} = -\left(j\omega / \omega_o\right)v_{o2}; v_{o4} = -v_{o2} \tag{7.27}$$

With this design, equation (7.27) suggests equal amplitudes for all four outputs, while the phase shift has already been given in the above discussion. It may be repeated that v_{o1} lags v_{o2}, v_{o2} lags v_{o3}, and v_{o3} lags v_{o4} by 90° in each case. The circuit can be further extended for realizing current outputs, by employing additional Z stages within the used DVCCs. The use of both Z+ and Z− stages within the DVCCs would provide current outputs at high impedance nodes. The readers are encouraged to undertake this as a problem for further study. The circuit can be modified for realizing an electronically tuned oscillator by employing tuneable versions of DVCCs. The resistively terminated X terminals provide easy compatibility of the circuit for the purpose. The circuit can thus be made resistor-less as well. It is yet another research problem which may invoke interest to the readers pursuing an advanced programme in the subject area.

Practice Exercise 7.10: *Design the circuit of Figure 7.8 using 20 pF capacitors for a frequency of 5 MHz. If the intrinsic X-terminal resistances of DVCCs are 70 Ω and the parasitic capacitances at the Y and Z nodes are 0.1 pF, calculate the error in the designed frequency of oscillation.*

Practice Exercise 7.11: *Assuming non-ideal DVCCs with finite current and voltage transfer gains (α and β respectively), obtain the frequency of oscillation for the circuit of Figure 7.8.*

7.5 ELECTRONICALLY TUNEABLE OSCILLATOR CIRCUITS

The oscillators discussed so far in the preceding sections are all based on the active-RC approach of realization. The active-RC oscillators, like any other active-RC network, cannot be classified as electronically tuneable networks, although it may be argued that the replacement of passive resistor by active means can provide tuning to such active-RC circuits. This is one of the standard methods to impart tuning to an active-RC circuit, but with limited tuning range. Another class of circuits with an inherent ability to vary a certain circuit parameter is more useful for realizing electronically tuneable circuits. This section is devoted to the study of such electronically tuneable oscillator circuits. The readers are already familiar with the available current-mode building blocks which fall under the tuneable category. We have been using such blocks in the treatment given thus far. The current controlled current conveyors (CCCIIs) have been often used for the purpose in this book so far. The CCCIIs are a favourable choice for realizing electronically tuneable oscillators. The first circuit to be studied herein is based on the active

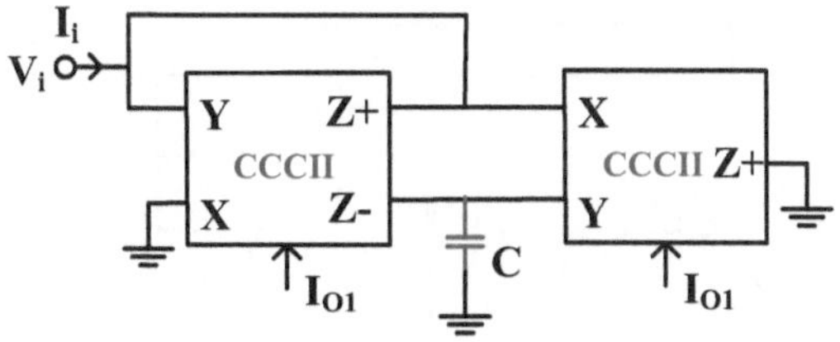

Figure 7.9 Electronically tuneable inductor simulator circuit [39].

inductor realized using CCCIIs and a capacitor. The circuit of active inductor is shown in Figure 7.9 [39].

The circuit can be analysed for its input impedance (Z_{in}). The following expression is found for the circuit's input impedance.

$$Z_{in} = \frac{sR_{x1}R_{x2}C_1}{1+(R_{x2}-R_{x1})C_1} \tag{7.28}$$

Equation (7.28) can be interpreted as follows. For the special case of $R_{x1} = R_{x2} = R_x$, the input impedance is inductive, as given below.

$$Z_{in} = sR_{x1}R_{x2}C_1 \tag{7.29}$$

Equation (7.29) suggests the effective inductance realized as $L_{eq} = R_{x1}R_{x2}C_1$. However, the expression of equation (7.28), without applying matching, simulates an inductor in shunt with a resistor of value $R_{x1}R_{x2}/(R_{x2} - R_{x1})$, which is equivalent to a parallel combination of a positive and a negative resistor. Therefore, the circuit of Figure 7.9 actually simulates an inductor in shunt with a combination of parallel positive and a negative resistor. The circuit is electronically tuneable. The variation of bias currents of CCCIIs results in varying intrinsic X-terminal resistances of CCCIIs. This provides a design clue to realize an oscillator using the popular RLC resonator technique. It may be noted that this technique depends on an RLC resonator with a negative resistance in shunt. Thus, the circuit of Figure 7.9 requires an additional capacitor to complete the oscillator topology. The resulting circuit of electronically tuneable oscillator is shown in Figure 7.10 [39]. The oscillator's characteristic equation is given as below.

$$s^2 + s\left(\frac{1}{R_{x2}C_2} - \frac{1}{R_{x1}C_2}\right) + \frac{1}{R_{x1}R_{x2}C_1C_2} = 0 \tag{7.30}$$

The FO and CO can be found from the above equation (7.30) and are given as below.

$$\omega_o = \sqrt{\frac{1}{R_{x1}R_{x2}C_1C_2}}; R_{x1} \leq R_{x2} \tag{7.31}$$

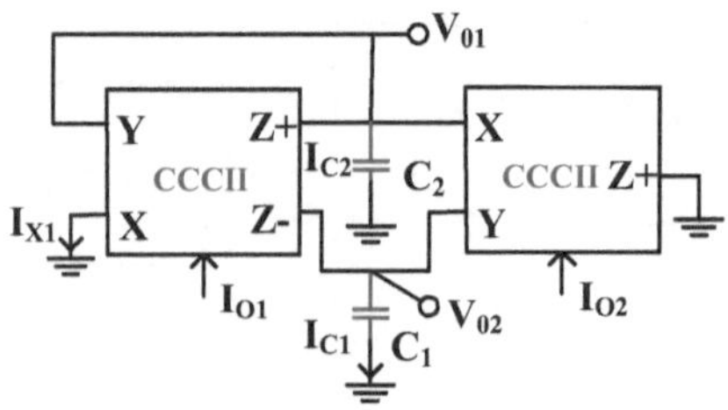

Figure 7.10 Electronically tuneable quadrature oscillator circuit [39].

The frequency and condition of oscillation as given in equation (7.31) suggests that the two CCCIIs are to be biased with equal currents. The oscillation frequency can be electronically controlled by the bias currents of CCCIIs. It is also clear that independent control over FO and CO is not possible in the circuit. The circuit can provide three current and two voltage outputs. The readers are encouraged to obtain the phase relationship between the various outputs. It may be mentioned that the three current outputs are phase shifted by progressive 90°, and the two voltage outputs are also quadrature in nature. The circuit can be designed for a given frequency of oscillation. As an example, if it is desired to use the bipolar version of CCCII, the intrinsic *X*-terminal resistance is $R_x = V_T/2I_o$, where I_o is the bias current of CCCII. For the circuit in question, there are two CCCIIs used in it. The design begins with assuming two capacitor values. If the capacitors are chosen as 20 pF each, and the design frequency is given as 2 MHz, then equation (7.31) can be used for finding the values of intrinsic resistances. The necessary bias currents (I_{o1} and I_{o2}) are next found, once the R_{x1} and R_{x2} values are known. Theoretically, the two bias currents are equal, but from the CO expression, it is required to adjust one of the two currents for sustained oscillation. This is only evident once the circuit is tested in laboratory conditions. For instance, under simulation-based testing, the condition of oscillation demands R_{x1} to be slightly less than R_{x2}, which means that I_{o1} is slightly more than I_{o2}. Therefore, one of the two currents is adjusted until oscillations are sustained. The design of the oscillator circuit in Figure 7.9 is a bit tricky in the sense that FO is not independent of CO. However, the FO tuning is only possible with matched bias currents, such that the CO is also maintained. The circuit design using CMOS CCCII would be similar except the expression for R_x, which is inversely proportional to the square root of the bias current, as already given in earlier chapters. The circuit outputs are useful only when the impedance level at the sensing nodes are appropriate. If this is not appropriate, the generator circuit would not provide the desired signal to the circuit being driven by it. For instance, if the circuit is being used as a carrier generator in a communication system, it is desired that the current outputs are available at high impedance nodes and voltage outputs are at low impedance nodes. Looking at the circuit of Figure 7.10, it is clear that none of the outputs enjoys this feature. The waveforms being generated are sinusoidal with defined phase shifts, but either not accessible at independent nodes or not at proper impedance levels. The actual sensing of signals generated requires additional sensing elements in the form of current and voltage followers. Some of these issues are cleverly addressed by using CCCIIs with multiple Z stages, especially for sensing the current outputs. This does not increase the CCCII count, rather only additional Z stages to be implemented within the existing CCCIIs. However, the access to voltage outputs may actually increase the active element count, because a voltage follower action requires the use of a current conveyor, exploiting the voltage conveying action of the element. The readers are encouraged to realize such a circuit with independent outputs with appropriate impedance levels. The circuit with high impedance current outputs can be found in existing literature as another oscillator with these features [40].

The problem of tuning the frequency of oscillation independent of the condition is solved in the next circuit to be studied in this section. It is also based on CCCIIs and capacitors. An electronically tuneable oscillator circuit employing three CCCIIs and two capacitors is shown in Figure 7.11 [41]. The oscillator's characteristic equation is expressed as below.

$$s^2 + s\left(\frac{1}{R_{x2}C_1} - \frac{1}{R_{x3}C_1}\right) + \frac{1}{R_{x1}R_{x2}C_1C_2} = 0 \qquad (7.32)$$

Equation (7.32) is different from the characteristic equation of the previous circuit of Figure 7.9. It is quite clear that the use of three CCCIIs provides the feature of independent control, which was missing in the previous circuit. After studying a number of oscillator circuits, it is now quite evident that the frequency of oscillation for the circuit of Figure 7.11 depends on the intrinsic resistances of two CCCIIs, marked with bias currents I_{o1} and I_{o2}. The condition of oscillation for the oscillator depends on I_{o2} and I_{o3}. This means that the FO can be tuned independent of CO by varying the bias current I_{o1}. Similarly, the CO can be set by adjusting I_{o3}, without affecting the FO. The readers are encouraged to write the expressions for FO and CO of the circuit. The electronically tuneable oscillator provides three current outputs, which are progressively shifted in phase by 90°.

Practice Exercise 7.12: *Obtain the phase relationship between the three outputs of Figure 7.11.*

Practice Exercise 7.13: *Design the oscillator circuit of Figure 7.11 using 10 pF capacitors for a varying FO from 1 MHz to 4 MHz using (i) bipolar CCCIIs; (ii) CMOS CCCIIs.*

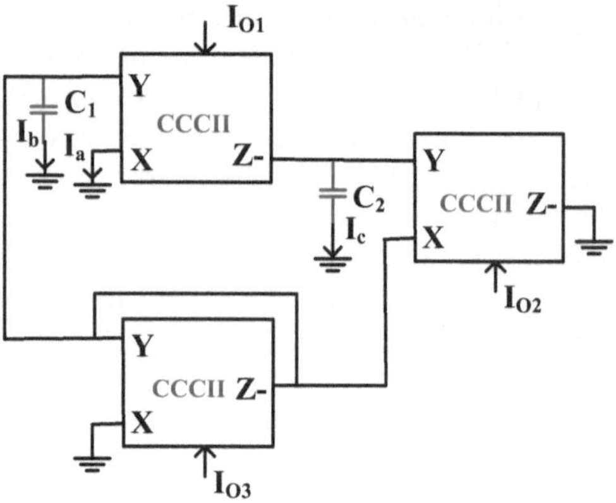

Figure 7.11 Another electronically tuneable quadrature oscillator [41].

The oscillator of Figure 7.11 also falls under the quadrature oscillator category. The circuit with grounded capacitors seems a good choice for integration. However, this advantage is lost once the outputs are to be accessed through capacitors. This is a practical problem which needs to be overcome like some of the other circuits already discussed previously. The additional current followers need to be employed for sensing the outputs at high impedance, while retaining the advantage of grounded capacitors. The solution to this problem lies in employing CCCIIs with multiple Z output stages. The marked outputs can all be accessed by incorporating additional Z stages. Two of the outputs are very obvious for this modification. The CCCII with bias current I_{o1}, when augmented with one $Z+$ and one $Z-$ additional stage realized the output shown across the X terminal and capacitor (C_2), respectively. The third output, which is across C_1, requires some analog design trick! The capacitor current (across C_1) can be easily expressed in terms of the Z-terminal currents of CCCIIs with bias currents I_{o3} and I_{o2}. More specifically, $I_{C1} = I_{Z3} + I_{Z2}$, which means that incorporation on an additional $Z+$ stage with the two CCCIIs (with bias currents I_{o3} and I_{o2}), and connecting these two stages together, provides the output current across capacitor C_1. Therefore, a simple modification allows all the three outputs to be available at the desired high impedance nodes. As another extension to the circuit, fourth output can further be generated by additional $Z-$ stages in two of the CCCIIs. Which of the two CCCIIs would provide a fourth output that is the inverted version of the current across C_1? How is the fourth output related to the other two outputs? Some of these questions are left for the readers to dwell upon.

7.6 CIRCUITS' REALIZATION USING ICs

The oscillators based on the active-RC realization approach are often built using off-the-shelf ICs. The operational amplifier-based oscillators have already become a part of a large number of text books available for undergraduate students of Electronics Engineering. The availability of such circuits provides a convincing platform for undergraduate learners to build circuits in laboratory environment using commercially available ICs. The current-mode techniques are best targeted for actual implementation of circuits in modern IC technologies, but the circuits' compatibility to available chips is an added asset. This is true for oscillators designed using current-mode techniques. The current feedback amplifier has already been introduced in earlier chapters. It has also been emphasized that the CFOA can be realized using commercial ICs. One such IC which has been referred to throughout so far is the AD844 chip. The IC realizes a CCII with an additional voltage follower. The oscillator circuit discussed in this chapter are all suited for being realized using such an IC. The CCII-based oscillators can be realized using a single chip for CCII+, and two chips for CCII–. Similarly, the DVCC-based circuits require three such ICs for one DVCC. Therefore, active-RC designs of current-mode oscillators are easy to test using AD-844 chips, which are commercially available [42, 43]. As an example of the circuit's easy compatibility to AD-844 IC, the circuit presented in Figure 7.3 is used for the illustration. The circuit is shown realized using three ICs, as also mentioned in the related section earlier. The AD-844-based circuit is shown in Figure 7.12.

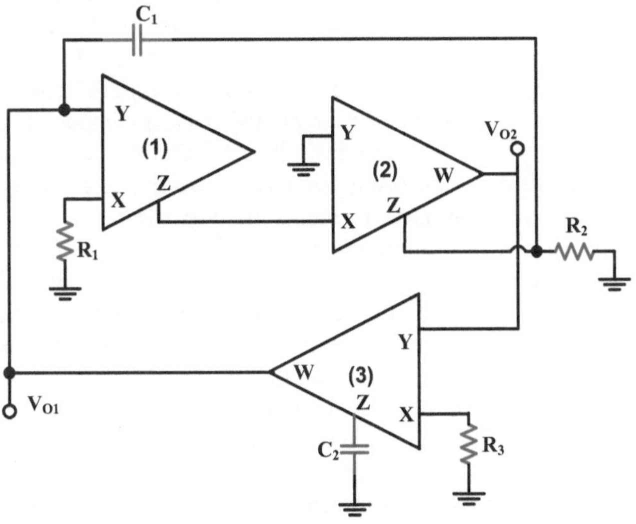

Figure 7.12 Example of AD-844 based quadrature oscillator.

The analysis of the circuit is carried out to obtain the FO and CO as given below. It may be noted that these expressions are same as for the circuit of Figure 7.3.

$$\omega_o = \sqrt{\frac{1}{R_1R_3C_1C_2}};R_2C_1 \leq R_3C_2 \tag{7.33}$$

From equation (7.33) it is evident that the circuit can be designed for a given frequency of oscillation. Let us choose capacitors readily available in an academic institute's electronics lab of value 100 pF. If the desired frequency of oscillation is 0.5 MHz, then the resistor values need to be calculated. The readers may go ahead with the design and assemble the oscillator circuit for testing using AD-844 ICs. In the next step, the frequency of oscillation may be varied from 0.5 MHz to 2.5 MHz, by varying R_1, which may be taken as a variable resistor. Both the outputs can be observed from available oscilloscopes. The phase shift between the two output waveforms is to be measured manually from the oscilloscope. The desired phase shift for the circuit is actually 90°. The X-Y mode of oscilloscope can be used for plotting the Lissajous pattern, which for this case has to be a circle, signifying the quadrature relation between the two outputs. The observed frequency of oscillation may be compared with the designed value. Once the error in frequency is known, the actual values of the passive elements need to be measured for the frequency achievable from the used elements. The error calculations are now to be repeated using the actual measured passive elements' values. The next exercise is related to finding the causes for this discrepancy in the designed and observed frequency of oscillation. The non-ideal model of the IC being used comes to the rescue for these analyses. The AD-844 model exhibits finite parasitic associated with the X, Y, Z and W terminals. These are approximately modelled as R_x for intrinsic X-terminal resistance, R_y, R_z and R_w as respective terminals' parasitic resistances, and C_y and C_z as respective terminals' parasitic capacitances. Once their actual values are known from data sheets, their effect on circuit performance is easy to assess. The oscillator circuit's parasitic analysis is to be performed, where these parasitic elements are actually reflected. Therefore, the causes of error can be found. In order to reduce the error in frequency of oscillation, some design guidelines are now discussed, which would be helpful to the readers. The choice of external passive elements is crucial for the accuracy of obtained frequency of oscillation. For the given circuit (Figure 7.12), the design guidelines are now mentioned. The resistors R_1 and R_3 are connected at the X terminal of the IC, which would introduce finite R_x in each of the two cases. Thus, these two resistors are to be large as compared to R_x. This means the choice of these two resistors in few kilo-ohms or more, so that the effect of finite R_x is minimized. The resistor R_2 is connected at the

Z terminal of one of the two ICs. The choice of this resistor is dictated by the Z-terminal resistance (R_z). Since R_3 appears in parallel with R_z, it needs to be much smaller than R_z, so as not to affect the circuit performance. Thus, R_3 can also be in few kilo-ohms and preferably not very high. One of the two capacitors in the circuit is floating, while the other one is grounded and connected at the Z terminal of the IC. This would be affected by the parasitic C_z and hence needs to be larger than the expected parasitic at the Z terminal. A good choice would be in tens of picofarads. Therefore, the other capacitor can be chosen accordingly in the same range. These guidelines would be helpful in reducing the errors in frequency of oscillation.

Rather than actually present the solution, this discussion is given as a problem, which would be appealing to a broader group of readers. The answers are not difficult but hopefully would keep the readers interested in actually exploring them. Besides, this further provides an opportunity to undertake such problems at various levels of programs being pursued by interested readers. In fact, the other circuits presented in the chapter can further be designed using such ICs and actually built in laboratories as mini-project problems. The complete chip implementation may be seen as a higher level problem for industry professionals and chip designers. To present a glimpse of the typical output of a quadrature oscillator built using ICs, Figure 7.13 is given. The two outputs seen in the figure exhibit the quadrature relationship.

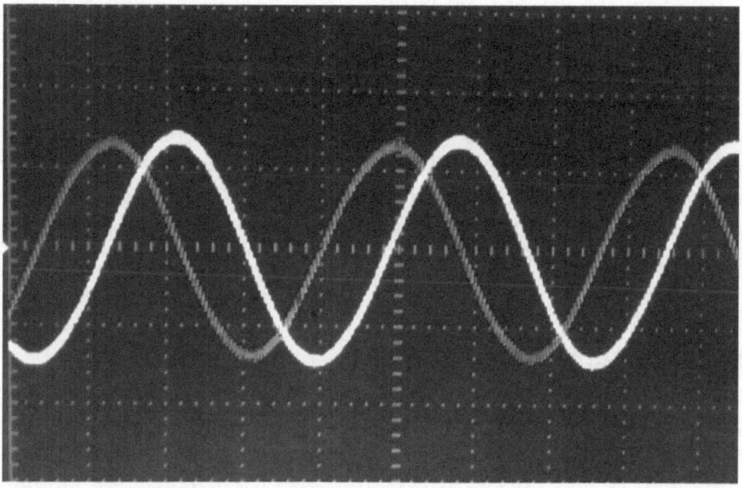

Figure 7.13 Typical quadrature oscillator outputs.

7.7 APPLICATIONS IN COMMUNICATION CIRCUITS

The applications of waveform generators in design and testing of electronic systems are numerous. The types of waveform generators covered so far in the treatment of current-mode techniques are sinusoidal generators. There are other varieties of signal generators often used in practice. The square wave, triangular wave, sawtooth wave, arbitrary wave are a few examples of other waveforms encountered in electronics and communication engineering. The current conveyors-based waveform generators are now common in open literature. The square waveform is one of the standards for clock and control-based applications in computing devices and systems. The use of the sawtooth waveform is well known in oscilloscopes. The various waveforms also form carrier of information in communication systems. Various modulated waveforms, such as amplitude shift keying and binary phase shift keying waveshapes, were shown in the earlier chapter. The pulse width modulated signal is also another application in communication circuitry. An interesting example of communication-based application of current-mode circuits is worth noting. It is based on the extended application of multiphase sinusoidal generator using DVCCs. It was earlier shown that the phase-shifted sinusoidal signals are of interest in communication and instrumentation. The appropriate selection of the signals so generated can yield various different waveforms of significance in communication systems. The design of MOS-based switching network can allow generation of different modulators [44]. An eight-phase oscillator has been used in ref [44] for realizing multiple PSK systems. The reported advance includes a QPSK, $\pi/4$ PSK and an 8-PSK system designed using DVCCs. The complete system is based on the use of CMOS DVCCs and MOS switches and hence is ideal for CMOS implementation with low supply voltage operation and all the advantages of current-mode techniques. Another recent advance using current-mode building blocks reports a square and triangular wave generator using a relatively complicated current-mode block [45]. The interesting part of the advance reported in [45] is the experimental verification of the idea by commercially available current-mode ICs. Many such recent developments can be explored by interested readers for advancing their know-how on the topic.

7.8 SUMMARIZED CONCLUSION

The general introduction to waveform generation and various methods for realizing such circuits was discussed. The current-mode techniques for sinusoidal oscillators was covered in detail. Several current-mode circuits for active-RC oscillators and electronically tuneable active-C oscillators were studied. CCIIs, DVCCs and CCCIIs were used to present the various theoretical and practical aspects of such circuits. The electronically tuneable oscillators designed using current controlled current conveyors provided feasible solution to the oscillators' realization problem in CMOS technology. The types of oscillators included were mainly quadrature and multiphase, with both voltage as well as current outputs. The possibility of testing the circuits in laboratory environment was discussed with a few examples. The simulation and experimental examples were included for the benefit of readers interested in pursuing the topic as part of research studies.

Chapter 8

Configurable analog blocks

This chapter is concerned with a relatively advanced concept in analog circuit design. The concept of programmable analog arrays is introduced in this chapter. These are referred to as field programmable analog arrays, and abbreviated as FPAAs. The current-mode techniques are used to present some interesting configurations, which are suitable for realizing programmable analog arrays. The usefulness of various current conveyors in realizing configurable analog blocks is shown, which forms the heart of FPAAs. The DVCC is mainly used for the purpose of illustration. Some simulations results are given to invoke interest amongst readers.

8.1 THE FPAA CONCEPT

The design of analog circuits and systems had been quite different from the design of digital circuits and systems. It is different from the digital circuits in the sense that a lot of the features that digital circuits enjoy by virtue of their simplicity were not present in analog circuits. These features have their origin in shrinking device dimensions and lower power supply operation, which reduced the size of digital circuits, while increasing their packing density along with reduced power consumption. The main parameters of interest in digital circuits are power and delay, which is often quoted as the power-delay product. On the other hand, the level of device shrinking for analog circuits in neither possible nor desirable from the perspective of meeting challenging specifications. Thus, higher packing density and lower power supply operation in analog design is a challenge. The specifications to be fulfilled in analog circuit design are not few; rather, it is a long list of parameters. Besides the power and delay, several other parameters like bandwidth, slew-rate, impedance levels, gain, noise, etc., are to be met in an optimal way. A large number of parameters to be fulfilled denotes design trade-offs in terms of these specifications. The digital design benefitted from the repeated use of identical blocks to realize digital systems. The hierarchical approach along with the design automation helped the digital design much more, leading to the complex memory and processor designs. The

DOI: 10.1201/9781003403111-8

need to modify/program circuits provided further motivation to the development of field programmable gate arrays. However, analog design mostly relies on assorted building blocks to realize an analog sub-system or system. The idea of hierarchical approach to design systems slowly found its way to analog design. The repeated use of inverters, amplifier, filters, transconductors, etc., led to the concept of field programmable analog arrays (FPAAs). The use of identical analog blocks for system design was the main motivation. The second motivation was to program/configure the design. The ease of analog design and reducing design time were other features being explored in the concept. The design flexibility in analog circuits and systems were also being searched in the new concept. The growing number of design techniques provided encouraging solutions in the area. Thus, FPAAs found some attention by the analog designers. The techniques available to explore the concept were mainly of two types. The continuous-time field programmable analog arrays and discrete-time filed programmable analog arrays thus started to become a reality. The continuous-time FPAAs were designed using operational amplifier and passive elements. The tuneable versions found the favour of operational transconductance amplifiers and capacitors as basic building blocks. The discrete-time FPAAs relied on the use of operational amplifiers, switches and capacitors, mainly designed using CMOS technology. In both the techniques, the requirements of inputting and outputting, the need of switching and programming were common. Another essential component for FPAA concept was the use of a cell/block with functional versatility, which was called the configurable analog block, abbreviated as CAB. The CABs designed using current-mode techniques can be studied further from various references on the topic, which are available in literature [46–48]. The following section takes the readers to the deliberations of this block.

8.2 NEED OF CONFIGURABLE ANALOG BLOCKS (CABs)

The configurable analog block (CAB) is often referred to as the heart of field programmable analog arrays. It is a network of active and passive elements arranged to perform distinct electronic functions, with operational flexibility. The CABs are employed to design FPAAs. These are often arranged in a matrix form, with each CAB programmed for a specific function. The overall system application depends on the connectivity of individual CABs and their individual functionality. The CAB is often designed using operational amplifiers, operational transconductance amplifiers, resistors, capacitors and switches. The advent of current-mode techniques has resulted in CABs based on current conveyor of various types. The CCII-RC-based CABs provide active-RC solutions, while the CCCII and capacitors-based CABs offer tuneable realizations. The other types of current conveyors are equally useful for designing CABs. These are DVCC, DXCCII, EXCCII for non-tuneable realization, and CCCDBAs and EXCCCIIs for tuneable ones. It is best desired to design CABs with tuneable analog building blocks like CCCIIs, EXCCCIIs and CCCDBAs. The circuit complexity of a CAB often dictates the functional complexity of the system being designed using the FPAA. The functional versatility of the CAB decides its usability to design complete analog systems using the FPAA concept. A generalized CAB with simplicity may be of use for distinct system-level applications. A generalized CAB with greater functional versatility may find more diverse system-level applications. Besides the continuous-time approach, the CABs can be designed for discrete-time analog systems, with easy compatibility to digital circuits. The discrete-time CABs using CMOS current conveyors are useable for such FPAA-based systems. Some of the simple configurable analog blocks are to be next considered in the section to follow.

8.3 CURRENT-MODE ANALOG CELL/BLOCK

The first configurable analog block to be presented is based on the DVCC. It is realized using a single DVCC and two impedances. The circuit topology is shown in Figure 8.1. The analysis is performed using the DVCC defining equations already given in earlier chapters. The two impedances are connected respectively at the X and $Z-$ terminals. The topology shows two nodes for inputting the voltage signals. The output voltage expression is given below.

$$V_o = \frac{-Z_1 V_{IN1} + Z_2 V_{IN2}}{Z_1 + Z_2} \tag{8.1}$$

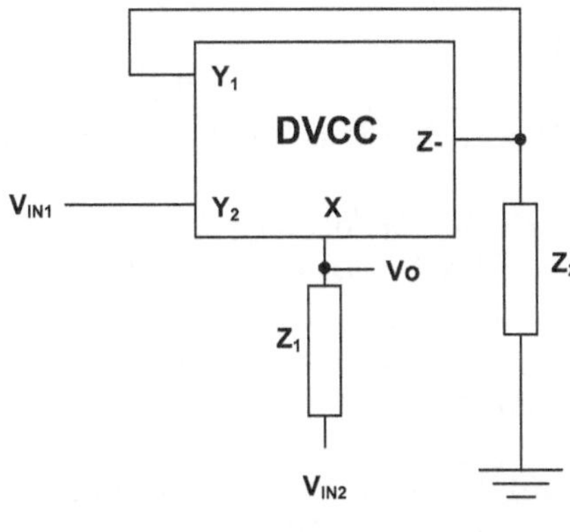

Figure 8.1 CAB using a single DVCC.

Table 8.1 Various analog functions realized

S. No.	V_{in1}	V_{in2}	Z_1	Z_2	*Transfer function*	*Function realized*
1.	0	V_{IN}	R	C	$\frac{1/RC}{s+1/RC}$	Low-pass filter
2.	V_{IN}	0	R	C	$\frac{-s}{s+1/RC}$	High-pass filter
3.	V_{IN}	V_{IN}	R	C	$-\frac{s-1/RC}{s+1/RC}$	All-pass filter
4.	0	V_{IN}	R_1	R_2	$\frac{R_2}{R_1+R_2}$	Voltage attenuator
5.	0	V_{IN}	R	R	0.5	Signal divide by two

The CAB can be appropriately designed for a function by selecting the node at which the input voltage is to be connected. Several possible circuit functions can be realized from the CAB. These are listed in Table 8.1. It is evident that the CAB performs useful simple electronic functions encountered in analog signal processing.

The specialization of the numerator of this transfer function yields low-pass, high-pass and all-pass filters, signal attenuators and signal divide by functions. It is to be noted that there is also no need for any component-matching conditions or inverting type voltage input signals to realize all the functions. The circuit of low-pass filter is realized when $V_{IN1} = 0$ and $V_{IN2} = V_{IN}$ in Figure 8.1. The choice of two impedances is as Z_1 as resistive and Z_2 as capacitive. The transfer function of the filter is given as below.

$$\frac{V_o}{V_{IN}} = \frac{1/RC}{s + 1/RC} \tag{8.2}$$

From equation (8.2), it is clear that a low-pass filter function is realized.

Practice Exercise 8.1: *Design the low-pass filter function (equation 8.2) for a pole-frequency of 1 MHz, assuming 10 pF capacitor.*

Practice Exercise 8.2: *The circuit diagram of Figure 8.2 using CAB is wrongly drawn using the Z+ terminal, instead of Z− terminal. What is the transfer function realized?*

The circuit of first order high-pass filter can be obtained, when Z_1 is chosen as resistive R, Z_2 as capacitive (1/*s*C), $V_{IN1} = V_{IN}$ and $V_{IN2} = 0$, as shown

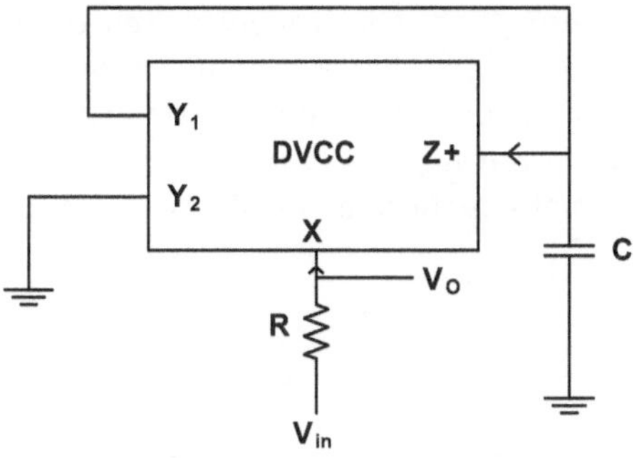

Figure 8.2 Circuit using CAB for analysis.

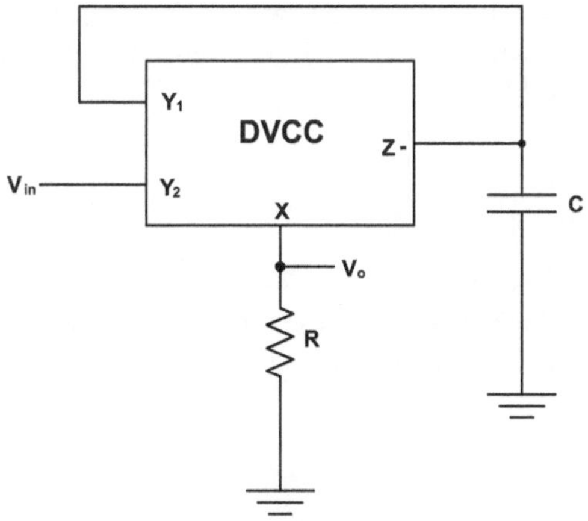

Figure 8.3 Circuit diagram of first order high-pass filter.

in Figure 8.3. The transfer function of the first order high-pass filter is given below as equation (8.3).

$$\frac{V_o}{V_{IN}} = -\frac{s}{s + \frac{1}{RC}} \tag{8.3}$$

Practice Exercise 8.3: *How can a band-pass filter be realized using two CABs using low-pass and high-pass functions?*

Practice Exercise 8.4: *The circuit of Figure 8.3 is modified so as to include a resistor (R_o) in shunt with the capacitor. Analyse the circuit so obtained for its transfer function.*

By assuming Z_1 as R_1 and Z_2 as R_2 in CAB topology of Figure 8.1, the circuit for voltage signal attenuator can be obtained. The output voltage is given below as equation (8.4). The readers are encouraged to obtain the circuit resulting from the above choice as an interesting exercise. The resulting circuit may then be designed for different attenuation, as another simple exercise. For instance, if the value of two resistors is selected as 2 kΩ each, the circuit provides an attenuation factor of 0.5.

$$V_O = \frac{R_2}{R_1 + R_2} V_{IN} \tag{8.4}$$

Practice Exercise 8.5: *With the help of Table 8.1, realize a high-pass filter with adjustable passband gain.*

DVCC-based magnitude divide by two circuits is obtained by assuming $Z_1 = Z_2 = R$, as shown in Figure 8.4. The output voltage of the circuit is given in equation (8.5).

$$V_O = \frac{V_{IN}}{2} \tag{8.5}$$

In order to study the exact behaviour of the CAB topology, it is necessary to perform the non-ideal analysis of the topology. By taking into account the non-idealities of a DVCC, the relationship of the terminal voltages and current can be rewritten as:

$$\begin{bmatrix} Vx \\ Iy1 \\ Iy2 \\ Iz- \end{bmatrix} = \begin{bmatrix} 0 & \beta 1 & -\beta 2 & 0 \\ 0 & 0 & 0 & 0 \\ 0 & 0 & 0 & 0 \\ -\alpha & 0 & 0 & 0 \end{bmatrix} \begin{bmatrix} Ix \\ Vy1 \\ Vy2 \\ Vz- \end{bmatrix} \tag{8.6}$$

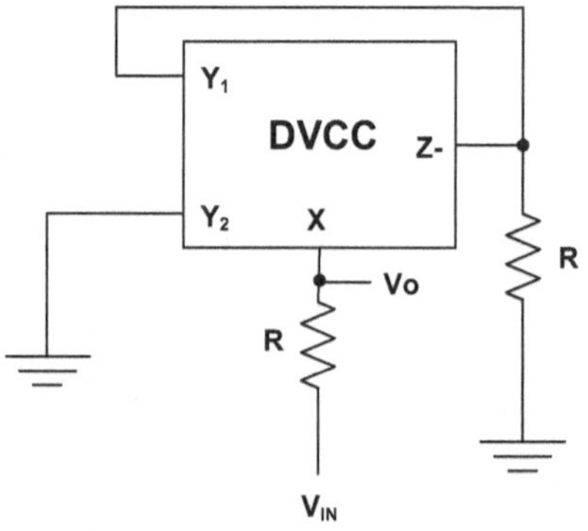

Figure 8.4 Magnitude divide by two circuits.

In equation (8.6), β_1 and β_2 are the voltage transfer gains from Y_1 and Y_2 to the X terminal, respectively, whereas α is the current transfer gain of the DVCC. These transfer gains differ from unity by the voltage and current tracking errors of the DVCC. Using the non-ideal description of the DVCC, the voltage transfer gains for the low-pass and high-pass filters are found as below.

$$\frac{V_{LP}}{V_{IN}} = \frac{\alpha\beta_1}{s + \alpha_2\beta_1/RC} \tag{8.7}$$

$$\frac{V_{HP}}{V_{IN}} = \frac{-\beta_2 s}{s + \alpha\beta_1/RC} \tag{8.8}$$

From equations (8.7) and (8.8), the cut-off frequency for low-pass and high-pass are given in equation (8.9).

$$\omega_{cl} = \frac{\alpha_2\beta_1}{RC}; \ \omega_{ch} = \frac{\alpha\beta_1}{RC} \tag{8.9}$$

The non-ideal transfer function of voltage attenuator is given below.

$$\frac{V_O}{V_{IN}} = \frac{\beta_1 R_2}{\beta_1 R_2 + \alpha R_1} \tag{8.10}$$

The effect of finite voltage and current transfer gains is clearly evident in the transfer function given in equation (8.10). The non-ideal transfer function of signal divide by 2 is given as equation (8.11).

$$\frac{V_O}{V_{IN}} = \frac{\beta_1}{1 + \beta_1} \tag{8.11}$$

The DVCC parasitic model is shown in Figure 8.5, which incorporates the various parasitic resistances and capacitances at the DVCC terminals. The DVCC-based CAB is now shown re-drawn in Figure 8.6. The circuit

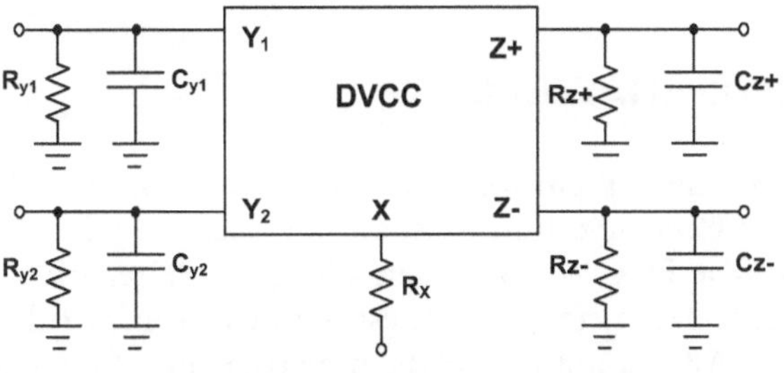

Figure 8.5 Symbol of DVCC showing parasitic elements.

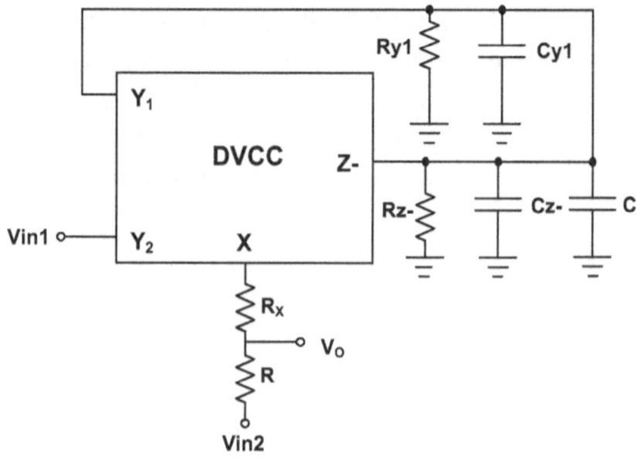

Figure 8.6 CAB topology showing parasitic elements.

incorporates the effects of parasitic elements. The readers are encouraged to analyse the circuit for its actual transfer function in the presence of parasitic elements. This would enable to study the actual behaviour of the CAB.

Many other useful current-mode analog blocks are available in open literature. The use of CCII and DVCC is very common in such blocks. The DVCC-based topology given in ref. [49] is one of the useful choices as CAB. The single DVCC-based topology is capable of realizing simple amplifier and single time constant networks. These are basic building blocks for higher order systems, as has already been studied in the previous chapters. Therefore, the topology of ref. [49] has been successfully employed to realize second order filters and oscillators. Another CAB that was recently designed is based on the use of EXCCII [50]. The topology uses a single EXCCII and three impedances, with a maximum of four passive elements. The CAB offers high input impedance for voltage-mode operation. The possible electronic functions performed by the CAB range from simple amplifiers, single time constant networks to second order filters. A total of 14 distinct functions are shown realized in that work. The CAB operates on a low power supply of only ±0.75 V. The topology has useful extensions for realizing other higher order analog systems. The readers are encouraged to refer to the details as given in ref. [50].

8.4 TUNING OF ANALOG CELLS

The CABs are an integral part of FPAAs. Besides being versatile in terms of realizing diverse electronic functions, an important requirement for CABs is their tuning. The FPAA environment is to provide real-time updating of the circuit/system parameters of interest with electronic or digital control. The active-RC CABs would not be the preferred choice for the purpose of tuneable systems. Although programmable resistor arrays (PRAs) are used in FPAAs which fulfil the need of varying resistive elements in active-RC-based CABs, programmable capacitor arrays (PCAs) are also employed in FPAAs, which provide another degree of tuning possibility with active-RC designs. The active-C technique-based CABs offer two degrees of tuning abilities: firstly, by the tuneable active building block and secondly by the PCAs. The use of tuneable analog building blocks like CCCII and EXCCCII is quite handy for designing CABs with inherent tuning feature. Similarly, other active building blocks like CCCDBA are also useful for the purpose. The DVCC with tuneable intrinsic X-terminal resistance also provides a choice to the analog designers. The current controlled DVCC is another option, which can be used for designing tuneable CABs. A mention of digital control was made above but needs more explanation. The use of a digital control word for varying a parameter in analog circuit design is known for a long time. The digital to analog converters are used in many applications, where the output of the converter acts on the analog circuitry, which is controlled by the digital input word of the converter. The auxiliary digital control circuit when used in the FPAA environment helps in a completely integrable and programmable system. Many of the tuning needs for such systems are still an open area for further work to be carried out in the future by prospective readers.

8.5 IC COMPATIBILITY ASPECTS

The configurable analog blocks are supposed to be realized using modern technologies. In this text, the technology being referred to is CMOS. A complete integration of the system is desired. However, the topic of CAB realization is still not a mature topic in analog circuit design. It is still confined to research groups or at the most to post-graduate levels of study. The undergraduate learners are to be introduced to this growing area from a different perspective. This perspective is connected to the compatibility of possible configurable analog blocks with the available ICs. Although not from current-mode techniques viewpoint, one example is worth mentioning. This is not within the scope of this book, yet may allow novice readers to appreciate its worth, because it involves the voltage operational amplifier. Most of the readers are well acquainted with a single opamp and two impedance-based popular topology. This topology is an excellent voltage-mode CAB, without the readers ever pondering on it from this perspective. The appropriate connections of the input voltage, ground terminal and impedance specialization are required to use this topology for distinct electronic function. It is easy to verify on the readers end that an amplifier of the inverting and non-inverting types, integrator (of lossy and lossless type), differentiator (lossy and lossless), voltage follower, Schmitt trigger and many other functions can be realized from it. The opamp-based voltage-mode topology is compatible with a single general purpose opamp IC. This example is given

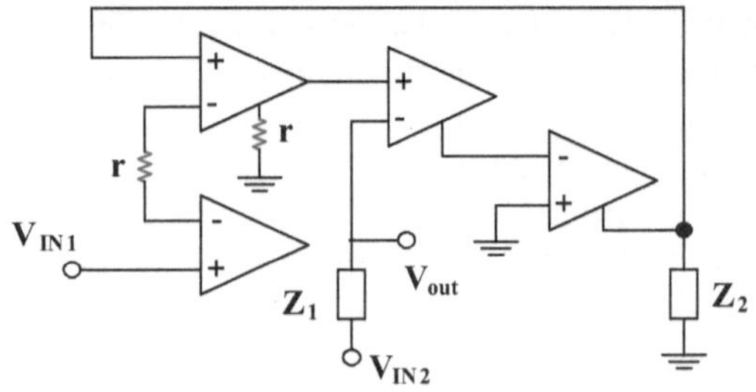

Figure 8.7 CAB topology realized using AD844 ICs.

for the undergraduate readers to actually verify the usefulness of traditional voltage opamp-based designs. For the present study, the current-mode active building blocks are of concern. Therefore, the CABs realized using CCII, DVCC or EXCCII are to be compatible with commercially available AD-844 ICs. The previous chapters have mentioned the possible realization of these building blocks. It is interesting to note that the DVCC-based CAB topology presented in earlier sections can be made as a starting point to breadboard the design using AD-844 ICs. The connection diagram for the purpose is shown in Figure 8.7. It may be noted that the output in the figure has been shown marked as V_{out}.

It may be noted that the CAB can be breadboarded using four ICs and two resistors, besides the two external impedances used in the topology. The DVCC type used in the CAB is of negative polarity; hence four ICs are required. For a DVCC of positive polarity three ICs are needed. Therefore, a topology using positive polarity may be a future thought for readers working in the area. It may be argued that use of four ICs for CAB experimental testing is an over-usage for the function. Although this is quite true, it is shown for the beginners to actually perform this as a laboratory exercise. Very compact CAB realizations using a single IC are also available. Single CFOA-based CABs have been shown to perform various analog functions. The CAB given in ref. [51] can be used as an integrator (both lossy and lossless), amplifier, etc., which find applications in the design of filters and oscillators. The oscillator example of the CAB is included in the work [51]. Useful extensions of the ideas presented so far are well reserved for future studies.

8.6 SPICE RESULTS AND CONCLUDING REMARKS

Before concluding the chapter, some real simulation results are presented for the readers interested to carry out further studies on the topic. The CAB introduced in Section 8.3 is used for the purpose. The detailed functionality of the CAB has already been presented. The DVCC-based topology is now simulated using the SPICE simulator. The low-pass, high-pass, all-pass and signal divide by two functions are actually obtained using simulation. The circuit for filtering functions is designed using a 10 kΩ resistor and a 100 pF capacitor, while for the signal divide by two, two equal resistors of 10 kΩ each are used. The DVCC is biased by a supply voltage of ±2.5V. The results are shown in Figure 8.8. The first figure (8.8a) shows the low-pass gain magnitude plot. Figure 8.8b is for the gain magnitude of high-pass function. The third figure (8.8c) shows the divide by two gain response. The last plot is for the all-pass filter, where Figure 8.8d shows the phase and gain magnitude response.

The concept of field programmable analog arrays was introduced for system design applications. The most important constituent of the FPAAs, a configurable analog block, was studied in detail. A simple CAB designed using a DVCC was presented with a detailed analysis of the parasitic and non-ideal behaviour of the topology. The usefulness of the CAB for various electronic functions was also shown. The tuning of CABs was emphasized, citing some recent works on the topic. The IC compatibility aspect was discussed towards the close of the chapter. It is expected that the chapter would provide a starting point for the readers to gain further understanding of the topic.

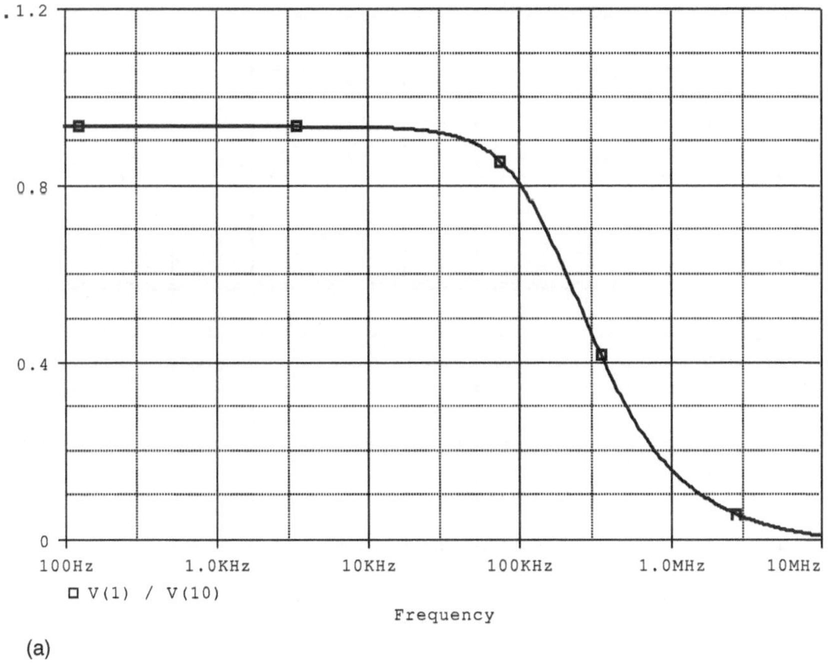

Figure 8.8a The gain magnitude response for low-pass filter.

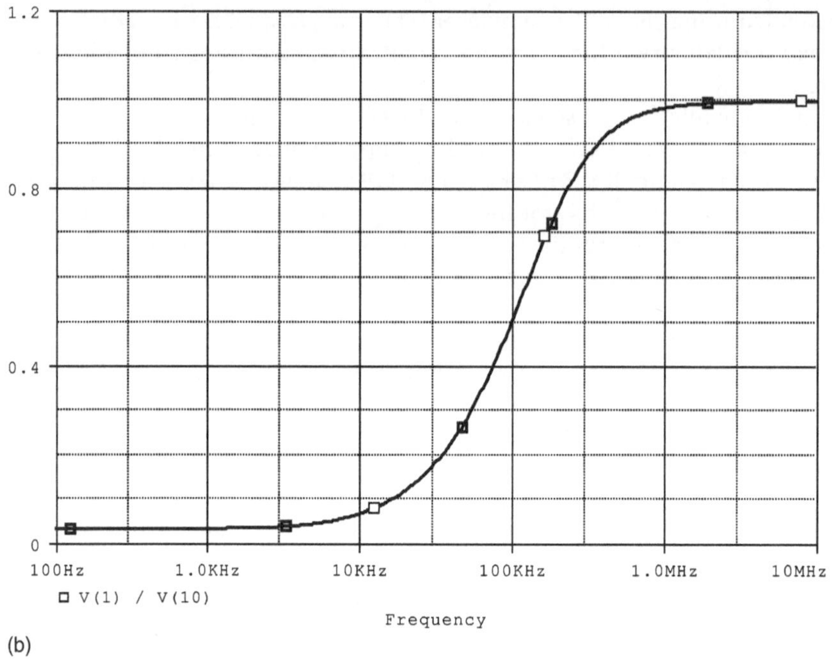

Figure 8.8b The gain magnitude response for high-pass filter.

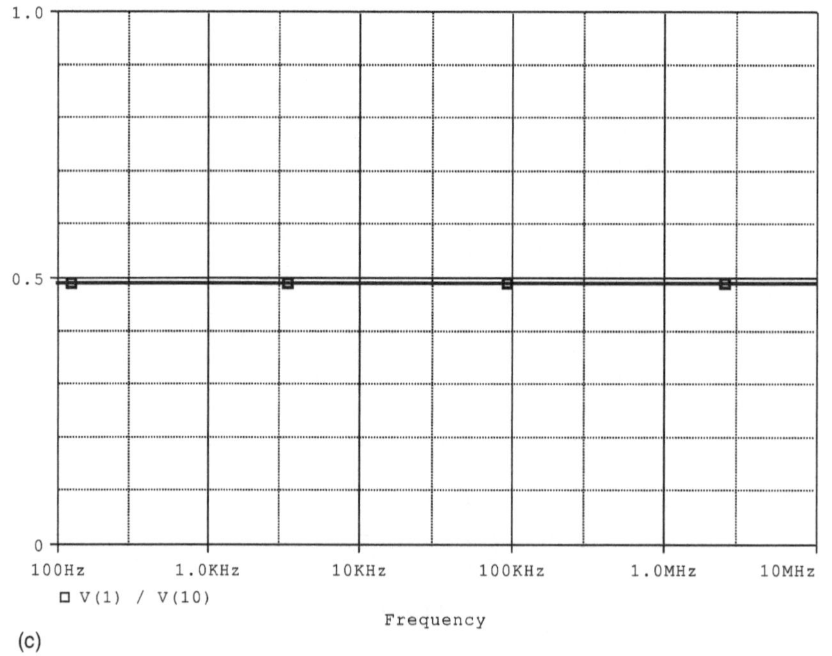

Figure 8.8c The gain magnitude response for signal divide by 2 circuit.

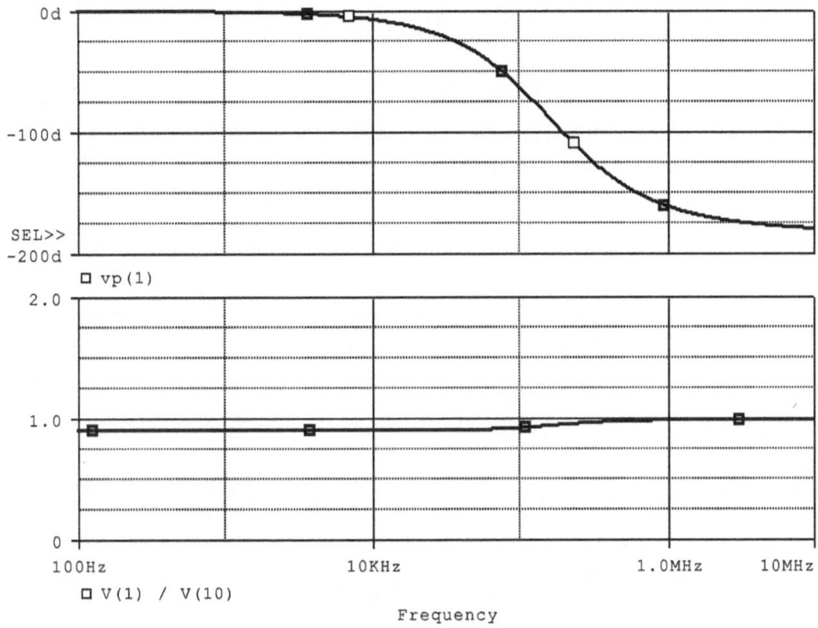

(d)

Figure 8.8d The phase and gain magnitude response for all-pass filter.

Chapter 9

Future of analog systems

This chapter is focussed on some of the futuristic concerns and contents which may provide definitive directions for designing analog circuits using current-mode techniques. The challenges for the analog circuits in a mixed signal environment are considered and some solutions suggested. The possibility of employing switched capacitor circuits using current-mode techniques is suggested for mixed signal design. The possibility of employing simpler analog blocks and dynamic threshold transistors with low voltage operation is explored. The advantages of migrating to current-mode designs are further strengthened by a novel circuit topology for neural applications. System-level design examples are included, which include qualitative discussion on digital modulation circuits, and new proposal on one of the important neural circuits. The chapter closes with general remarks on the future needs and trends.

9.1 CHALLENGES FOR MIXED SIGNAL ENVIRONMENT

The analog circuit design using current-mode approach offers many opportunities for mixed signal environments, which may be viewed as challenges for circuit designers. The contrasting requirements of digital and analog blocks have been earlier discussed in the previous chapters. The compatibility of analog designs with digital circuitry is a topic of continuous study. The various design techniques relaying on the current-mode approach may be considered for the present study. The challenges for mixed signal environments are numerous. The first one is the use of low-voltage-operated analog circuits with easy compatibility to digital circuits. The second challenge is to integrate successfully the analog blocks with digital blocks and even realize digital functions using analog building blocks. The third problem to be solved is to minimize the analog usage, by adopting analog sampled data circuits. Another problem to be addressed is to employ simpler analog blocks with an optimum transistor count. The other challenges are related to the speed of operation, frequency range of signals to be handled and the linearity aspects. The layout of circuits with minimum crosstalk, reduced

DOI: 10.1201/9781003403111-9

interference, protection of sensitive signal lines, need of differential signalling, symmetry considerations, etc., are also to be appropriately dealt with while designing such circuits for a mixed signal environment. The challenges mentioned above are best reserved as design layout considerations, not to be further elaborated. However, in the present context, the solutions to the other above-mentioned challenges are worth further study. These are elaborated below as interesting design options.

Some of the low-voltage-operated current-mode circuits are quite interesting [52–53]. One low voltage current conveyor using an inverter error amplifier can be found in literature, which operates on a supply voltage of ±0.5 V. The 20-transistors-based current conveyor has been used for designing an oscillator circuit for portable systems [52]. One low-voltage-operated digitally controlled fully differential current conveyor which also operates on ±0.5 V has been used for designing field programmable analog arrays. The tuneable amplifier and filters based on this block are reported in the work [53]. The dynamic threshold MOS transistor (DTMOS)-based current conveyor is used in a work for realizing an all-pass filter. The circuit operates on a supply voltage of only ±0.2 V and the power dissipation is also low, i.e. 0.53 μW [54]. The filter is quite suitable for realizing higher order analog circuits. Many such advances can be found in recent literature on the topic. Such low-voltage-operated circuits will find easy merger with digital blocks. The next type of circuits with easy compatibility with digital circuits and possible digital function realization using analog blocks has also been reported in open literature. One such advance based on analog building blocks reports digital function realization using extra-X current conveyor. The basic logic functions like OR, AND, NOR and NAND are shown realized in the work, which are based on the current-mode technique [55]. Each of the functions is based on current inputs and current output. The advance reported employs the basic current following topology of current conveyors. The reported idea may be useful for future systems, where digital functions may be realized using the analog blocks for the sake of modularity and in situations demanding control over an analog design parameter by digital means. The third type of circuits designed using current-mode techniques for future challenges pertain to the switched capacitor approach. Under this approach, some of the recent advances with a futuristic view are reported in literature. One such work uses current conveyors for realizing simple analog functions using the switched capacitor approach [56]. Simple functions are realized using current feedback opamps, which may find useful applications in analog sampled data signal processing. The extra-X current conveyor-based switched capacitor circuit for simple electronic functions is also given therein [56]. The other future challenges to be addressed are to reduce the transistor count for analog signal processing. The use of current-mode blocks with reduced transistor count has prompted the popularity of voltage and current follower-based realizations. Simple analog functions can be realized using this method. One such work is based on realizing filters using such blocks [57]. The low voltage operation and reduced transistor count will surely prove to be a boon for designs suited for mixed signal environments, which will lead to the design of fully current-mode systems.

9.2 MIGRATION TO CURRENT-MODE TECHNIQUES: A NOVEL CASE STUDY

After a broad coverage on current-mode techniques, it is a matter of time that these techniques would be soon accepted by the designers. The replacement of existing voltage-mode designs is a challenge because of the maturity of transitional approaches and their time-tested behaviours. However, it is to be realized that the migration to new techniques is the need of future electronic and communication circuits and systems. One example is illustrated herein, which presents a novel circuit idea. A novel current-mode circuit for realizing bipolar sigmoid activation function is proposed. The new circuit uses an extra-X current conveyor and two MOS transistors, as an inverter. The circuit offers advantageous features of accepting current input signals at low impedance nodes, providing output voltage as bipolar sigmoid activation function, operation at ±1.75 V supply, CMOS compatible structure and generality of varying threshold (reference) levels. The new circuit has potential applications in neural networks. The simulation results are given to verify the proposed circuit.

The applications of artificial neural networks have encompassed a very broad spectrum of daily life human activities of survival and luxuries, so much so that the listing of such applications itself is a voluminous task. The recent advancements in the area have focussed on the use of analog design techniques for realizing the various functions for the purpose. This is mainly due to the dependence of CMOS technology for reducing the power and area of the designed modules, and the possibility of configuring the analog blocks according to the desired specifications. A function of importance in the neural network design is sigmoidal function. It is basically used as activation function for learning complex decision functions. Sigmoid function may lie between two values, namely, 0 to 1 or –1 to 1, which are useful in probability prediction. The former variation is obtained with sigmoid function, while the latter variation is achieved with a tangent hyperbolic function. The proposed circuit for realizing an activation function with possible neural network applications is shown in Figure 9.1. It comprises a current-mode building block, namely an EXCCII along with two transistors, being used as an inverter. The circuit of the EXCCII (already given in earlier chapters) is repeated as Figure 9.2. The EXCCII as shown in Figure 9.2 is characterized by the following port relationship, already given in earlier chapters but repeated below.

$$i_Y = 0; v_{X1} = v_Y; v_{X2} = v_Y; i_{Z1} = i_{X1}; i_{Z2-} = -i_{X2} \tag{9.1}$$

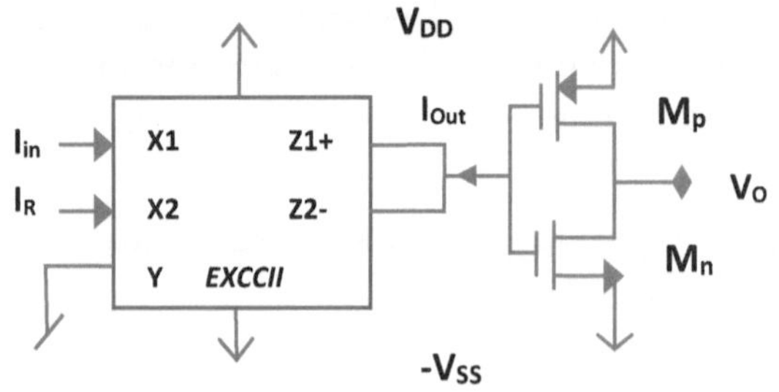

Figure 9.1 Proposed current-mode activation function circuit using EXCCII.

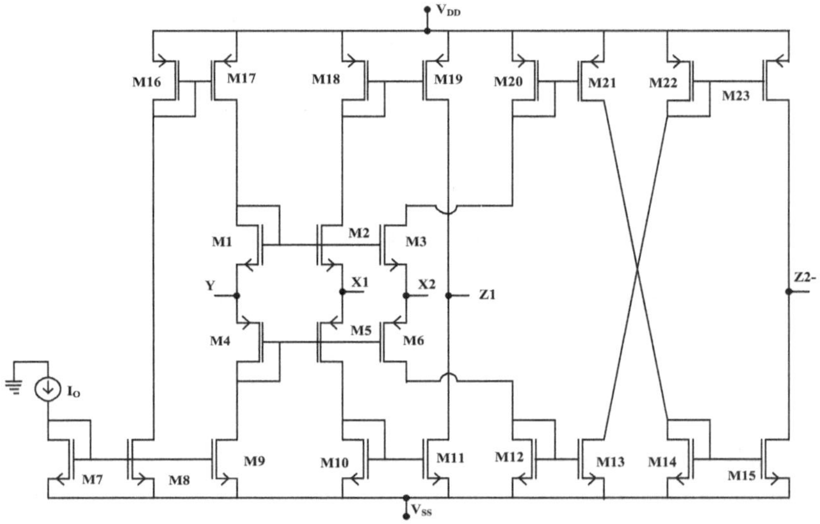

Figure 9.2 EXCCII circuitry used for designing the circuit of Figure 9.1.

Equation (9.1) needs some discussion for the sake of completeness. The input voltage terminal (Y) with ideally infinite input resistance, the other two current input terminals (X_1 and X_2) with ideally zero input resistance and the output current terminals (Z_1 and Z_{2-}) comprise an ideal EXCCII. The voltage follower action between the Y and the two X terminals along with the current follower action between X_1 to Z_1 and between X_2 and Z_{2-} are of positive and negative polarity, respectively. The proposed circuit of Figure 9.1 shows a grounded Y terminal, enabling the EXCCII to operate as two current followers: one with +1 current transfer gain and the other with −1 current transfer gain. The simplified expression for the topology of Figure 9.1 is given below.

$$I_{\text{out}} = I_{Z1} - I_{Z2} = I_{\text{in}} - I_R \tag{9.2}$$

Coming to the proposed circuit of Figure 9.1, the EXCCII is used for differencing the currents inputted at two X terminals, out of which one is the reference current. It may be noted that the current inputs are at low impedance nodes of EXCCII, namely the X terminals. The output current of the circuit (I_{out}) is given in equation (9.2), for Figure 9.1. It may be noted that the use of uppercase notations (equation 9.2) are in conformity with the proposed circuit application for activation function realization. The voltage at the gates (V_G) of inverter forming transistors can be expressed as below. It may be noted that the output resistance is modelled as R_{out}, which comprises the shorted Z stages' output resistance and the gates' resistance. Since the gates' resistance is ideally infinite, the effective R_{out} is parallel combination of R_{z1} and R_{z2}.

$$V_G = I_{out} R_{out} \tag{9.3}$$

$$R_{out} = \frac{R_{z1} R_{z2}}{R_{z1} + R_{z2}} \tag{9.4}$$

The output current I_{out} is positive for $I_{in} > I_R$, and the available voltage at the gates of the MOS transistors decreases, enabling the M_P to conduct. The circuit output (V_o) tends to reach the V_{DD}. On the other hand, for $I_{in} < I_R$, the available voltage at shorted gates tends to rise. This forces the M_n to conduct the current. The circuit output drops to negative supply value ($-V_{SS}$). The nature of output voltage is a replica of the sigmoid function. The output swing between positive and negative extremes (−1, 1) makes it a tangent hyperbolic function, which is also called the bipolar sigmoid function. The general form of bipolar sigmoid function is expressed as below.

$$y = \frac{e^x - e^{-x}}{e^x + e^{-x}} \tag{9.5}$$

The function (equation 9.5) has extensive applications in neural networks as activation function. The advantage of this function over the sigmoid function is due to its bipolar nature and zero-centred property. The scope herein is to propose a new circuit for one of the various possible activation functions.

The proposed activation circuit with current inputs and voltage output is verified through simulations. The CMOS implementation of the EXCCII is simulated using 0.25 μm process parameters with a supply voltage of ±1.75 V. The bias current of EXCCII (I_o) is fixed at 90 μA. The circuit is inputted with the DC currents (I_{in} and I_R). The inputs and output of the circuit of Figure 9.1 are shown in Figure 9.3. The output is clearly a bipolar sigmoid function, which is also referred to as the tangent hyperbolic function. The generality of the proposed circuit is further emphasized by varying the reference current, and the results obtained are shown in Figure 9.4, for positive values of the reference currents. The next set of results shown in Figure 9.5

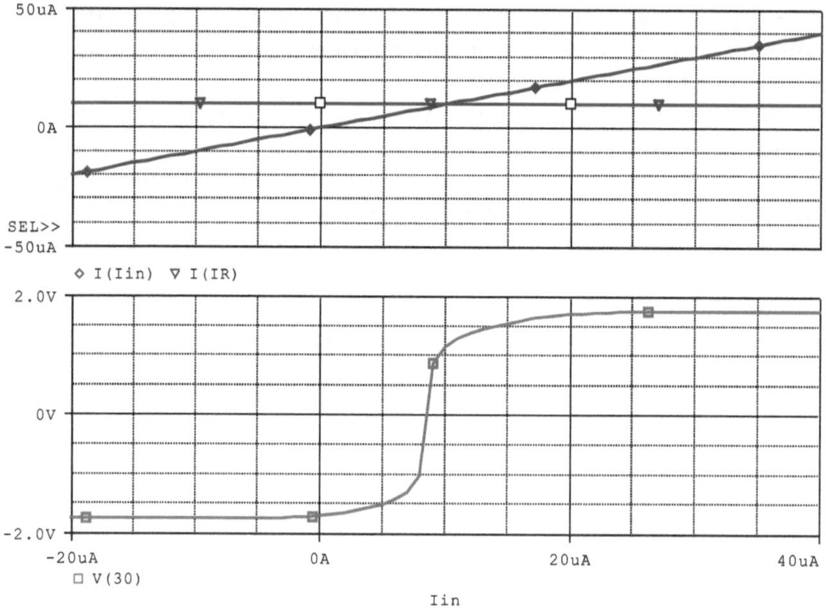

Figure 9.3 The inputs and the circuit output [V(30)] of Figure 9.1.

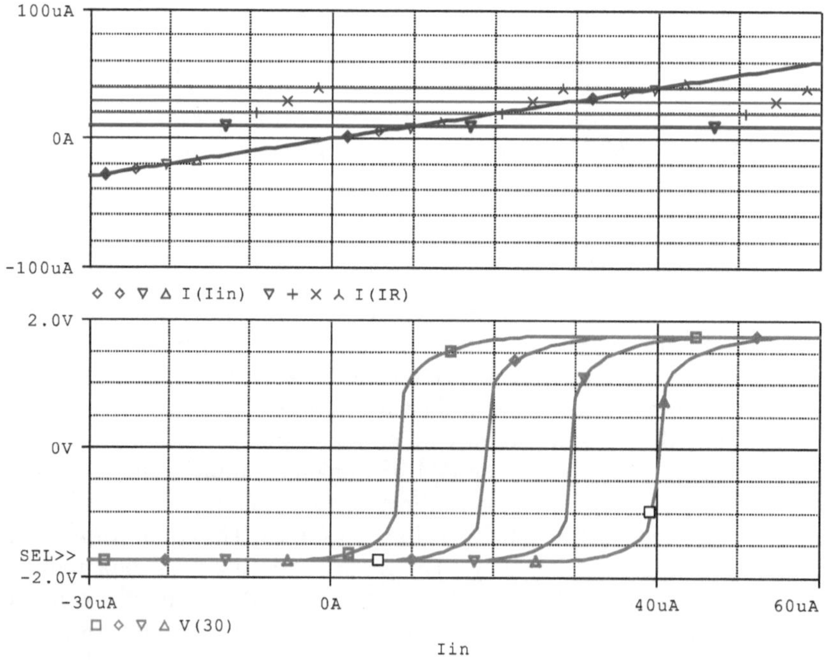

Figure 9.4 Input (I_{in}), varying reference current (I_R as:10, 20, 30, 40 μA) and output.

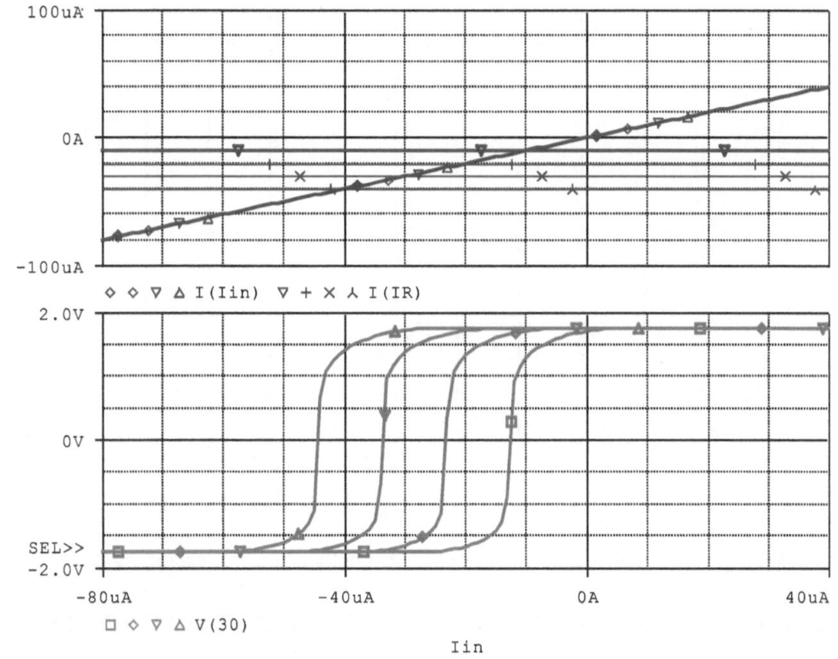

Figure 9.5 Input (I_{in}), varying reference current (I_R as: −10, −20, −30, −`40 μA) and output.

is plotted for negative values of reference currents. It is quite evident from Figures 9.4 and 9.5 that the proposed circuit is generalized and works well for different values of reference currents

The variation of temperature on the output function is studied for 0–50°C, in step of 10°C. The resulting plot is shown Figure 9.6, which shows overlapping curves for varying temperature. The power dissipation of the proposed circuit is found as 3 mW.

The proposed activation function is realized using a single EXCCII and two MOS transistors, working as an inverter. The input currents appear at low impedance nodes, which avoids the loading problem. The simplicity of the approach, the use of current-mode building block, operation at ±1.75 V supply and flexibility of its functionality for varying values of threshold levels make the new circuit a useful option for future neural network applications. The simulation results are given in support of the proposed theory.

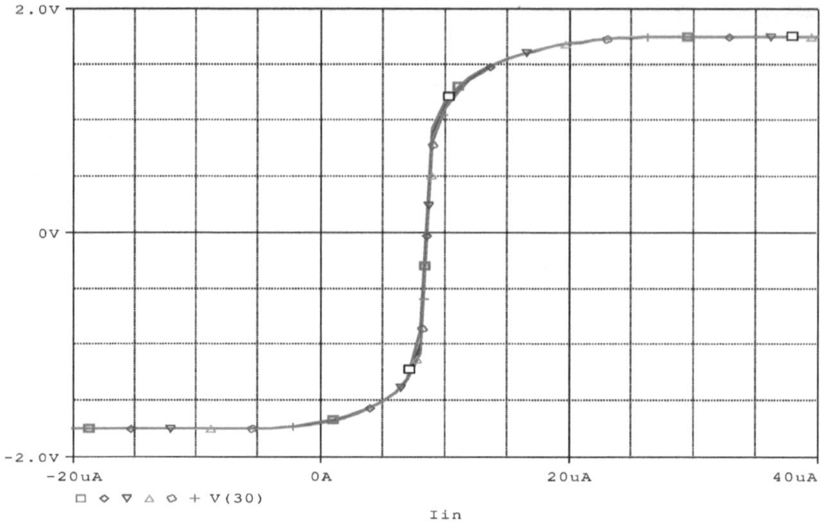

Figure 9.6 Temperature variation effect seen negligible on overlapping curves.

9.3 SYSTEM DESIGN USING CURRENT-MODE TECHNIQUES

Throughout the previous chapters and in the current chapter, various electronic circuits have been shown using the current-mode techniques. The functionality of the presented circuits in the book encompass simple current-mode analog interfaces, amplifiers, filters, rectifiers, comparators, oscillators and multipliers, which are essential building blocks for system design. The other essential building blocks covered include simple logic functions, summing/differencing units, integrators, and differentiators, which also find applications in system design. The coverage so far also presented a few examples of systems being realized using current-mode techniques. The waveform generation chapter has already introduced a multiple PSK system design using DVCCs. Another example worth considering herein is of another digital modulation scheme using current-mode techniques. The generation of ASK, PSK and FSK is a genuine system design application, which has been solved using current-mode techniques. The EXCCII has been used for the purpose. The simple scheme is fully compatible in CMOS technology. The use of a low voltage active building block for this scheme has made it a prospective future option. This advance is available in ref. [58]. The scheme presented therein uses a total of 33 MOS transistors. The use of 180 nm parameters is made, with a supply voltage of ±0.9 V for verifying the validity of the scheme. The operation of the modulator in current mode is shown, where the output is obtained at high impedance, which is ideal for cascading without additional buffers. The maximum carrier signal frequency of the reported advance is given as 50 MHz. The results for three basic digital modulations, namely ASK, PSK and FSK, are shown, both in simulation as well as experimental form. For the experimental purpose, the available current-mode ICs are used. The complete circuit layout is also given in this research paper, which further underlines its feasibility aspects. The detailed parasitic study is performed to study the real behaviour of this scheme. This is a perfect example of a system design application using the current-mode technique. Another example of system design using current-mode techniques is based on a recent published research [59]. The four quadrant analog multiplier finds extensive applications in the design of communication systems. The reported work is based on the dual *X* current conveyor and employs a total of 22 MOS transistors for the purposed. The circuit provides output current at high impedance node and accepts voltage signals as inputs. The circuit is simulated using 180 nm process parameters and operates on a supply voltage of ±0.9 V. The circuit exhibits a bandwidth of around 19 GHz. The design of a modulator, squarer and frequency doubler is shown in the work, with prospective future system design applications. Another example of system design using current-mode techniques is the ASK synchronous detector reported in one recent advance on the topic [60]. The use of EXCCII is made in the scheme for the purpose. The basic blocks of the scheme, namely the rectifier, low-pass filter and comparator are all realized using the current-mode technique. The implementation of the scheme is done and the results are given therein [60]. One work which is quite worthy of discussion herein is an interesting shelving equalizer, with low-shelf, high-shelf and band-shelf characteristics obtained using the popular EXCCII as active building block [61]. The simple structure with a single EXCCII, three resistors and one capacitor is used in different configurations for shelving applications. The low-shelf and high-shelf circuits are presented therein [61]. The band-shelf function is also shown realized. The simplicity of the topology operable at a supply voltage of ±1.25 V and useable up to 30 MHz makes the development suited for system design applications. Many such system design applications can be further found in recent works, and many more are expected to be discovered in future studies. The author urges the readers to undertake this topic, with the instigation given so far in this book.

9.4 RESEARCH INITIATIVES AND FUTURE DIRECTIONS

The treatment in this book has now reached a stage where it is important to point out certain missing gaps to be fulfilled in the future. The coverage so far has mainly been confined to certain current-mode building blocks. There are several other current-mode building blocks available, and many other may evolve in future. Some of these are voltage differencing gain amplifier, universal current conveyor, current conveyor transconductance amplifier, gain controlled current conveyor and buffered output extra-X current conveyor. These blocks have not been fully explored in the reported literature.

Similarly, the extended technologies need to be explored for designing the current-mode building blocks. The use of carbon nanotube field effect transistor, bulk driven MOS transistors, fin-shaped field effect transistors, etc., is a possible future research direction for the researchers. The design of low voltage current-mode circuits is still a growing area, which needs more attention as a future initiative. There is a strong need to employ differential topologies for real benefits of the mixed signal design and better noise elimination aspects. The beneficiary areas of the techniques are biomedical signal processing, analog computing and neuromorphic analog circuits. The role of mem-elements realized using current-mode techniques would grow in the future applications. The academic and research groups in the area need to collaborate with industry and fabrication labs to realize the benefits of current-mode techniques. There is a strong need to integrate the complete systems for diverse applications. The analog circuit design is an area which will continue to grow and impact the future electronic and communication system design.

9.5 CONCLUDING REMARKS

This chapter presented some interesting examples and a novel case study of current-mode analog design techniques. The brief coverage does show a glimpse of realized dreams and many future dreams to be realized. The journey of analog circuit design using current-mode techniques has reached a close for now, which needs to be picked up by interested readers and researchers working in the area or being instigated by this reading. The luxuries of mankind mentioned in the Preface may grow with each new initiative in the direction. This is expected to keep the academicians and workers in the area busy, with the growing numbers of enthusiastic new academicians joining the exploration journey.

References

[1] P. E. Allen, D. R. Holberg, *CMOS Analog Circuit Design*. Oxford Press, 2011.

[2] B. Razavi, *Design of Analog CMOS Integrated Circuits*. Tata McGraw-Hill Education, 2002.

[3] C. Toumazou, F. John Lidgey, D. Haigh, *Analogue IC Design: The Current-Mode Approach*. Presbyterian Publishing Corp, 1990.

[4] K. C. Smith, A. S. Sedra, The current conveyor—A new circuit building block, *Proceedings of the IEEE*, vol. 56, pp. 1368–1369, 1968.

[5] A. S. Sedra, K. C. Smith, A second-generation current conveyor and its applications, *IEEE Transactions on Circuit Theory*, vol. 17, pp. 132–134, 1970.

[6] S. Maheshwari, R. Verma, Electronically tunable sinusoidal oscillator circuit, *Active and Passive Electronic Components*, vol. 2012, Article ID 719376, 6 pages, 2012.

[7] S. Maheshwari, D. Agrawal, High performance voltage-mode tunable all-pass section, *Journal of Circuits, Systems and Computers*, vol. 24, p. 1550080, 2015. doi: 10.1142/S0218126615500802.

[8] B. Chaturvedi, S. Maheshwari, Third-order quadrature oscillator circuit with current and voltage outputs, *International Scholarly Research Notices (ISRN)*, vol. 2013, Article ID 385062, 8 pages, 2013.

[9] A. Fabre, O. Saaid, F. Wiest, C. Boucheron, High frequency applications based on a new current controlled conveyor, *IEEE Transactions on Circuits and Systems I: Fundamental Theory and Applications*, vol. 43, pp. 82–91, 1996.

[10] S. Maheshwari, Current conveyor all-pass sections: brief review and novel solution, *The Scientific World Journal*, vol. 2013, Article ID 429391, 6 pages, 2013. doi: 10.1155/2013/429391.

[11] S. Maheshwari, Additional summing/difference amplifiers using CCCIIs, *Journal of Active Passive Electron Devices*, vol. 1, pp. 159–162, 2005.

[12] S. Maheshwari, High CMRR wide bandwidth instrumentation amplifier using current controlled conveyors, *International Journal of Electronics*, vol. 89, pp. 889–896, 2002.

[13] S. Maheshwari, Active-only current controlled summing/difference amplifiers using CCCIIs, *Active and Passive Electronic Components*, vol. 26, Article ID 467841, 4 pages, 2003.

[14] A. Toker, S. Ozcan, H. Kuntman, Supplementary all-pass sections with reduced number of passive elements using a single current conveyor, *International Journal of Electronics*, vol. 88, pp. 969–976, 2001.

[15] I. A. Khan, S. Maheshwari, Simple first order all-pass section using a single CCII, *International Journal of Electronics*, vol. 87, pp. 303–306, 2000.
[16] S. Maheshwari, New voltage and current-mode APS using current controlled conveyor, *International Journal of Electronics*, vol. 91, pp. 735–743, 2004.
[17] S. Maheshwari, I. A. Khan, Simple first-order translinear-C current-mode all-pass sections, *International Journal of Electronics*, vol. 90, pp. 79–85, 2003.
[18] S. Maheshwari, A new current-mode current-controlled all-pass section, *Journal of Circuits, Systems, and Computers*, vol. 16, pp. 181–189, 2007.
[19] S. Maheshwari, CMOS compatible single input five/six outputs first order filter circuits, *Journal of Active & Passive Electronic Devices*, vol. 3, pp. 125–133, 2008.
[20] S. Maheshwari, B. Chaturvedi, Additional high input low output impedance analog networks, *Active and Passive Electronic Components*, vol. 2013, Article ID 574925, 9 pages, 2013. doi: 10.1155/2013/574925.
[21] W. Tangsrirat, On the realization of first-order current-mode AP/HP filter, *Radioengineering*, vol. 22, pp. 1007–1015, 2013.
[22] J. W. Horng, High input impedance first-order allpass, highpass and low-pass filters with grounded capacitor using single DVCC, *Indian Journal of Engineering & Materials Science*, vol. 17, pp. 175–178, 2010.
[23] B. Chaturvedi, A. Kumar, J. Mohan, Low voltage operated current-mode first-order universal filter and sinusoidal oscillator suitable for signal processing applications, *AEU-International Journal of Electronics and Communications*, vol. 99, pp. 110–118, 2019.
[24] H. P. Chen, Single CCII based voltage mode universal filter, *Analog Integrated Circuits and Signal Processing*, vol. 62, pp. 259–262, 2010.
[25] J. W. Horng, C. C. Tsai, M. H. Lee, Novel universal voltage mode biquad filter with three inputs and one output using only two current conveyors, *International Journal of Electronics*, vol. 80, pp. 543–546, 1996.
[26] B. Chaturvedi, S. Maheshwari, Current mode biquad filter with minimum component count, *Active and Passive Electronic Components*, vol. 2011, Article ID 391642, 7 pages, 2011. doi: 10.1155/2011/391642.
[27] S. Maheshwari, I. A. Khan, Novel cascadable current mode translinear-C universal filter, *Active and Passive Electronic Components*, vol. 27, pp. 215–218, 2004.
[28] J. Mohan, S. Maheshwari, Supplementary high-input impedance voltage-mode universal biquadratic filter using DVCCs, *Modelling and Simulation in Engineering*, vol. 2012, Article ID 184829, 6 pages, 2012.
[29] M. A. Ibrahim, S. Minaei, H. Kuntman, A 22.5 MHz current-mode KHN-biquad using differential voltage current conveyor and grounded passive elements, *AEU-International Journal of Electronics and Communications*, vol. 59, pp. 311–318, 2005.
[30] W. Tangsrirat, Current-tunable current-mode multifunction filter based on dual-output current-controlled conveyors, *AEU-International Journal of Electronics and Communications*, vol. 61, pp. 528–533, 2007.
[31] S. Maheshwari, Analogue signal processing applications using a new circuit topology, *IET Circuits, Devices & Systems*, vol. 3, pp. 106–115, 2009.
[32] S. Maheshwari, I. A. Khan, High performance versatile Translinear-C universal filter, *Journal of Active and Passive Electronic Devices*, vol. 1, pp. 41–51, 2005.

[33] M. Jahan, S. Maheshwari, Tuning approach applied for new band-pass filter circuit, *Journal of Active and Passive Electronic Devices*, vol 16, no. 1, pp. 71–81, 2021.
[34] S. Maheshwari, Tuning approach for first-order filters and new current-mode circuit example, *IET: Circuits Devices and Systems*, vol. 12, pp. 478–485, 2018.
[35] S. Maheshwari, Current controlled precision rectifier circuits, *Journal of Circuits, Systems, and Computers*, vol. 16, pp. 129–138, 2007.
[36] S. Maheshwari, Novel cascadable current-mode first order all-pass sections, *International Journal of Electronics*, vol. 94, no. 11, pp. 995–1003, 2007.
[37] S. Maheshwari, Sinusoidal generator with π/4-shifted four/eight voltage outputs employing four grounded components and two/six active elements, *Active and Passive Electronic Components*, vol. 2014, Article ID 480590, 7 pages, 2014.
[38] S. Maheshwari, Voltage-mode four-phase sinusoidal generator and its useful extensions, *Active and Passive Electronic Components*, vol. 2013, Article ID 685939, 8 pages, 2013.
[39] S. Maheshwari, I. A. Khan, Novel voltage/current-mode translinear-C quadrature oscillator, *Journal of Active and Passive Electronic Devices*, vol. 2, pp. 235–239, 2007.
[40] S. Maheshwari, High performance mixed-mode quadrature oscillator, *Journal of Active and Passive Electronic Devices*, vol. 2, pp. 223–226, 2007.
[41] S. Maheshwari, Electronically tunable quadrature oscillator using translinear conveyors and grounded capacitors, *Active and Passive Electronic Components*, vol. 26, Article ID 197109, 4 pages, 2003.
[42] S. Maheshwari, M. S. Ansari, Catalogue of realizations for DXCCII using commercially available ICs and applications, *Radioengineering*, vol. 21, pp. 281–289, 2012.
[43] S. Maheshwari, Realization approach for sinusoidal signal generation and circuit with easy control, *Journal of Circuits, Systems, and Computers*, vol. 29, no. 2, p. 2050031, 2020.
[44] I. Ali, S. Maheshwari, Multiple PSK modulator design using current mode building blocks, *2017 International conference on Multimedia Signal Processing and Communication Technologies, (IMPACT)*, 2017, pp. 262–266. doi: 10.1109/MSPCT.2017.8364017.
[45] A. Kumar, B. Chaturvedi, S. Maheshwari, A fully electronically controlled Schmitt trigger and duty cycle modulated waveform generator, *International Journal of Circuit Theory and Applications*, vol. 45, no. 12, pp. 2157–2180, 2017.
[46] O. O. Fares, M. T. Abuelma'atti, Configurable analogue building blocks for field-programmable analogue arrays, *International Journal of Electronics*, vol. 95, pp. 1009–1028, 2008.
[47] A. H. Madian, S. A. Mahmoud, A. M. Soliman, Configurable analog block based on CFOA and its application, *WSEAS Transactions on Electronics*, vol. 5, pp. 220–225, 2008.
[48] S. Maheshwari, Ten transistors based configurable analog cell, *2013 International Conference on Multimedia, Signal Processing and Communication Technologies*, Aligarh, India, 2013, pp. 246–249.

[49] S. Maheshwari, Analogue signal processing applications using a new circuit topology, *IET Circuits Devices Systems*, vol. 3, no. 3, pp. 106–115, 2009.
[50] S. Maheshwari, Configurable analog block using current mode approach, *Journal of Circuits, Systems, and Computers*, vol. 31, no. 2, 2250037, 18 pages, 2022.
[51] K. Maheshwari, S. Maheshwari, P. Yadav, Design of Configurable Analog Block-Based Oscillator and Possible Applications. In: Sharma, D. K., Balas, V. E., Son, L. H., Sharma, R., Cengiz, K. (eds.) *Micro-Electronics and Telecommunication Engineering*. Lecture Notes in Networks and Systems, vol. 106. Springer, Singapore. doi: 10.1007/978-981-15-2329-8_63.
[52] Y.-S. Hwang et al., A low voltage current conveyor using inverter-based error amplifier and its oscillator application, *IEICE Electronics Express*, vol. 10, no. 24, pp. 1–7, 2013.
[53] S. A. Mahmood, E. A. Soliman, Low voltage current conveyor-based field programmable analog array, *Journal of Circuits, Systems, and Computers*, vol. 20, no. 4, pp. 1677–1701, 2011.
[54] B. Singh, S. Maheshwari, All pass filter using DTMOS technique, *2020 International Conference on Smart Electronics and Communication (ICOSEC)*, 2020, pp. 1301–1305. doi: 10.1109/ICOSEC49089.2020.9215420.
[55] S. Maheshwari, Logic functions for mixed signal circuit design using analog block, *Australian Journal of Electrical and Electronics Engineering*, vol. 19, no. 3, pp. 300–305, 2022.
[56] S. Maheshwari, I. A. Khan, Switched-capacitor equivalent of some active-RC filters and a novel circuit, *Journal of Active and Passive Electronic Devices*, vol. 16, no. 1, pp. 33–49, 2021.
[57] A. Sharma, S. Maheshwari, Current follower based current mode filters, *2020 Third International Conference on Smart Systems and Inventive Technology (ICSSIT)*, 2020, pp. 632–636. doi: 10.1109/ICSSIT48917.2020.9214292.
[58] D. Agrawal, S. Maheshwari, Design and implementation of current mode circuit for digital modulation, *Integration, the VLSI Journal*, vol. 78, pp. 118–123, 2021.
[59] J. Rajpoot, S. Maheshwari, High performance four quadrant analog multiplier using DXCCII, *Circuits Systems and Signal Processing*, vol. 39, no. 1, pp. 54–64, 2020.
[60] S. Maheshwari, Detection of amplitude shift keying signals using current mode scheme, *International Journal of Electronics and Information Engineering*, vol. 11, no. 2, pp. 73–80, 2019.
[61] S. Maheshwari, Some analog filters of reduced complexity with shelving and multifunctional characteristics, *Journal of Circuits, Systems and Computers*, vol. 27, no. 10, p. 1850150, 2018.

Index